I0470038

Everything you

Never thought you'd

Need to Know

About Goats

I want to say a special Thank You to Sandra Miller, a great mother and wonderful friend, for helping edit this book. Thank you Mom!

Everything you Never thought you'd Need to Know about Goats

Copyright © 2013 Elwood Ranch - John & Melissa Foster
All rights reserved. No part of this of this book may be reproduced in any form or by any means without permission in writing from the authors.

Foreward

Where to start? … Goats are a passion of mine. They are wonderful as pets, and provide a natural resource for dairy products and meat for your family. They are also great land clearers of brush and weeds. All around I can think of no other animal I enjoy having more.

No matter what your reason is for wanting goats, you have two options – Jump In or Research beforehand. I recommend the latter but know many like myself who just jumped in and wished they had done the research first. I grew up with a lot of livestock/animals, so I wasn't completely uniformed about how to raise and take care of an animal. But goats are nothing like any other animal and if you aren't prepared, you'll find yourself in turmoil pretty fast.

I started raising my own goats, chickens, turkeys and rabbits in my early twenties. God, where has the time gone? When I first began my adventure, I thought fence, feed, water and *wahlah,* there you go. I was so wrong. Yes, I bought a book on goats, but did I really pay attention to it –no. I read through it twice and still I didn't follow any of the advice. Then, when I needed help I returned to the book and found it had a bunch of general nonsense - nothing that could really help me. I started my endless search for information on goat health and treatments. To my surprise it was very difficult to find the information I was looking for. In my mind, I thought there would be one website with all the information I would ever need, and unfortunately that was just not the case. When I finally thought I found good information, there was always another article or person disagreeing or saying to do it another way. I felt hopeless.

So now, after all my years of learning (twenty-so of them so far) and trying different things to see what really worked and what didn't, and what was true and what wasn't, going through the masses of information to determine what was false, what could be, and what actually was correct, I feel confident in my methods and knowledge to offer my advice to others. Maybe I can help someone starting out to not have to go through the learning curve I went through. As my mom always told me, if you can help just one person it is worth it. So here I am, about to begin this journey of writing a book with all the knowledge I have acquired in hopes of helping someone like you.

A little of due diligence, disclosure and CYA – All the information in this book has been gathered from my experience, knowledge, research and/or training. I am not a veterinarian, I am not certified in anything, and the information contained in this book is not to be construed as veterinarian advice or counsel. It is a documentary of sorts on my experience with goats. It is recommended to always consult a livestock veterinarian whenever a goat is ill. With that said, I hope you read further. You can also peruse our website at www.elwoodranch.com

Table of Contents

Chapter One – Before Getting Goats

Or what I wish I knew then –

If you start off right and develop a herd management program, your goats will be healthier. Review all the information here and then sit down and develop a plan for your herd before you even get a goat. Doing so will not only give you a good idea about what it is actually going to cost to raise goats, but it will help prevent many illnesses, disease, worms, and ill effects within your herd. Nothing is more important than a Great Herd Management Program. And yes, you can do this on a budget!

Goats are herd animals and cannot be raised alone without other goats. You must start out with at least two goats. Goats can be noisy, and this may become an issue with neighbors. Goats are highly addictive, be prepared to always want more. Goats can also be destructive. You have to prepare your facilities for their behavior.   I tell anyone wanting to start out raising goats to ask themselves these ten questions:

- How many goats do you think you will get?
- Do you know the difference between a browser and a grazer and which are goats?
- Do you know what a goat should eat and what they cannot eat?
- What are the different dietary requirements of a buck, pregnant doe, and kid?
- What type of a fence and shelter does a goat need?

- Do you have or can you set up quarantine areas?
- Is your land prepared for goats – zoning laws allow goats?
- Do you know the most common illnesses and treatments in goats?
- Do you have emergency care supplies on hand?
- Do you have a veterinarian available for goats?

Plan your herd arrangement based on your land. Goats need land that is dry – no wetlands, marshes, etc. Watch your land flow during heavy rain. Look at where the water flows and where it settles. <u>Goats hate water and don't like rain or puddles.</u> If you have enough land, you can fence in these areas as long as the major portion is high and dry. But, don't build shelter areas or huts in the wet areas.

Hilltops are a great space for goats. They love to climb and be where they can overlook their surroundings. If you don't have hills or slopes, you can build areas for them to climb and play on.

Forested areas or overgrown, undeveloped areas are great for goats. Goats are browsers and love to stand up and eat. Briars, wild berries, even rose bushes will all be eaten by a

goat. Just beware of any poisonous plants that may be on your property and remove them prior to getting goats.

All land will die out in the winter and you must be prepared to provide additional hay/feed for goats during these times. No browse means need for more hay which means more expense. The more pasture or browse area you have, the less expense there is for hay/feed costs. We have a lot of pasture area with good browse, so this cuts down on our hay costs.

Fencing is very important with goats. Your fence plans should protect the goat from getting out and keep predators from getting in. Goats are curious and natural browsers. They don't think about escaping, but if they see something that catches their eye, they go towards it and if there's a way they will get out. There's a saying if water flows through, so will a goat. It's not quite that bad, but for the most part they do get out of or through places you would never think they could. I know not everyone can put up 6 foot wooden panel fencing for their property, but your fencing has to be sturdy and small holed. Regular field fence will not keep a goat in and definitely will not keep predators out. So, what type of fence is best for a goat? What fence can a goat get out of? What type of animals are predators to a goat? Read more about fencing needs in the chapter on Herd Management.

The more shelters the better and they can be as inexpensive or as elaborate as you want or have the budget for. Plans for your shelter are part of your basic herd management program and setup should be based on the goat's needs, your safety practices, and health concerns. Goats need shelter to protect them from rain, wind and to feel safe. Shelters should be placed in areas based on your land as stated above. How should the shelter be built? What are the minimum requirements for a shelter? What type of flooring is available and what is best for your situation? Read more about shelters in the chapter on Herd Management.

There are many reasons why you need separate areas. Maintaining separate areas is a must for goat owners who plan to breed, sell, have both does and bucks, or bring in new goats. Again this can be simple and inexpensive or elaborate, depending on your budget.

Bucks and does have different dietary needs, and a pregnant doe needs a different diet than a non-pregnant doe. Also, if you choose to have bucks and does and want to plan your births, which I recommend, you need separate pen areas, shelters, etc. for both. You will also need a separate area for moms-to-be, for when you have weaned and are getting ready to sell kids, and for quarantining new goats or sick goats. Quarantine or isolation pens are a must in a good herd management program. What is a good plan for building separate areas/pens? Why is there a need for quarantine? Read more about this in the Herd Management chapter.

Goats require a lot of care when it comes to their nutrition. Improper feeding is the number one cause of illness within a goat herd, mostly because their rumen is very easily upset. Never make quick changes to a goat's diet. Goat's don't eat everything, nor do they eat tin cans. Goats are browsers and if fresh browse is not available this must be supplemented with good quality hay, feed, and supplements. In fact, even with good browse, hay is a mainstay in a goat's diet, especially with bucks. Grains/feeds are secondary and should be limited. What are the nutritional requirements of goats? What is the best feed plan? Read more about this in the chapter on Nutrition.

Health checkups and routine care is the second most important part of a well thought out herd management program. Every goat will need regular checkups and routine care for hooves, eyes, ears, mouth, gums, teeth, fur, body, fecal checks, general appearance, walk or gait, and temperatures. Your barnyard area and shelters should also be inspected, cleaned and repaired just as regularly as your goats. It is our belief that by following these guidelines, goats are healthier, their health can be maintained properly, and there is less illness and disease within a herd. Read more about health checkups and routine care in the Herd Management chapter.

Common Misconceptions

So, here are some things I have heard from wanna-be goat owners or new owners that although they are nice to think, don't usually work out the way you expect.

"I have (insert acres here) fenced with regular 4ft wire fencing and more open pasture with cow fencing."

Reality - 4 Ft regular fencing will work only if it is the No-Climb horse fence with the small 2×3 holes and only for your mini breeds. Cow fencing will never work. Full-size breeds need taller fencing and higher gauge wire. Goats love to browse and will easily fit their heads through, under or over fence that is not right for them. Some will even jump over the fence or crawl under. The wrong type fencing is not only unwise; it can be a health hazard for the herd if a goat gets its head stuck within the fencing.

"I have 3 acres and can get 20-30 goats an acre."

Reality – Full size goats should be no more than 7 per acre; Minis have a max of 10 per acre. Overcrowding leads to disease and health issues. Because goats are browsers they will eventually eat what you have and with more than the max number of goats per acre, the lesser/weaker goats will not feed properly.

"I can easily make a profit at this within a year or so."

Reality – Many goat owners never see a profit. It takes time, money, and effort to raise goats. If a goat comes down sick, it can wipe out your whole herd or cost an exorbitant amount of money to get back on track. Those who do their research prior to getting goats, and those who know their clientele and breed for the customer, could begin to see a profit within 3-4 years.

"I'll get some goats cheap or free and sell to show owners and make the money."

Reality – There's a reason those goats are cheap or free. They are most likely carriers of an illness, not great milkers, or already sick. It's best to spend the money upfront and purchase quality registered goats that have no health problems and good conformation. Goats cannot be shown without being registered and most must be dehorned – another expense.

"I can't afford to get two or more goats right now, so I think I'll just start with one goat."

Reality – Goats are herd animals. They don't do well alone or alone with other animals. If you can't afford to start with two goats, you probably can't afford to start goat herding. Don't just buy one goat. They'll get lonely, depressed and then sick.

"I have dogs so my one goat will be fine. He'll have friends."

Reality – Most dogs attack goats. At the very least they will chase your goats and cause them stress. Stress in goats leads to illness. Sick goats lead to more expense for treatment or death of your goat.

"I'm not looking to get rich, just a few extra hundred dollars a month."

Reality – I can almost 100% guarantee you will NOT get rich raising goats. Only a few breeds cycle year round, so there will not be monthly income. It takes time to build a clientele that will buy your goats and IMO you'll never get to the point where you have a steady monthly income. Don't expect to have kids and have buyers just waiting. To get to where you are making sustainable income comes from breeding **quality** not quantity. Quality for a wider group of buyers; Quantity will come as you build clientele and thereby more income.

"Well I'll just buy a couple of does and a buck and go from there."

Reality – Those who are new to goats should start small. Build your clientele before you even get goats. Get to know what is needed in your area. I would start out buying quality registered goats to begin with. That way you can do the 4H clubs, sell for show, and sell for meat/milk. Start small to see how you like goat ranching. It's not for everyone. Purchase two does already bred or breed them. Once you have decided this is what you want to do, I would invest in a great buck. He is the star of your show. It saves so much time, money, frustration, and eliminates a lot of disease issues by having your own closed herd – meaning you don't have to bring your does somewhere to be bred or bring in other goats to your does which offer less chance of contagions.

"I have a pasture area fenced in properly, so I just need my goats now."

Reality – Start up is a lot more than just fencing. You need shelters or a barn, kid areas, birthing areas, feeders, waterers, Start up costs are a lot. You should also stock up on feed, hay, (yes, you'll need feed and hay even on pasture) and medical supplies. If you are going to sell to 4H or for shows, you'll also need disbudding equipment and banding equipment.

"I heard people can sell their goats for XXX amount of money each."

Reality – Most goat owners never sell their goats for those extreme prices you see. It is a rarity and usually only in the show world after several generations of proven winners. Pricing and what they sell for are two different things. Yes you can advertise the higher prices, but really it all depends on where you are and who your clientele will be, and how much they are willing to pay. Again this is where research and buying quality stock from the beginning comes to play.

Here's some advice – I built my clientele even before I had any goats to sell by talking to people, finding out what they needed. I had people waiting to purchase before I got them because of my research. But it takes time and determination. It's not an overnight success and I'm not a big rancher. It's not something I could make a living at without other income. I'd say if you do the ground work now, and do it well, within a couple of years you can be making a profit and continue from there. But don't – DON'T just jump in thinking it will all come out in the wash and you'll have people wanting to buy your goats. Don't overprice your head either – yes it feels good to say I sold a goat for $300 but if that's the only goat you sell in a year what have you made? I undersold mine to begin with just to get the clientele started. Once they knew me, knew my grade (quality) I could sell for more and make the money. Service is also a key factor. I think I am the only one around my area who offers a guarantee for my goats, and I'm always willing to offer advice Free Of Charge, so that makes a big difference too. I also continue to monitor my market and switched breeds to fit the need around me, not because of how they looked or how cute they are.

If after reading this chapter you still feel goats are right for you, then ...Welcome to the World of Goats!

Chapter Two – Herd Management

A good herd management program starts with research and planning. It must take into consideration the number of goats you will start with and the possibility of increasing the size of the herd either through new purchases or kids born on site. The topics covered in this chapter include Barnyard Basics and Goat Management. We will discuss each of these topics in depth.

Barnyard Basics

This is the first preparation step for your herd management program and the most important. Barnyard Basics determines what your needs/wants are for goats. It covers the layout of your land, your water flow and flood areas, soil and plant composition, fencing requirements, shelter needs, pen and quarantine areas, placement of waterers and feed troughs, and your electrical and water needs.

So, why do you want to raise goats … as pets, for dairy, meat, to show, to sell, to clear land, or a combination of the above? What breed of goats are you thinking about raising? Once you decide why and what, you can prepare for the rest. But let me say, goats are addictive and you will most likely end up getting more than you planned for and change your reasons for keeping them. You will also need to determine how you are going to raise them. Are they going to have pasture to browse on – full time, part time, occasionally; or will they be raised dry-lotted without any natural browse.

The amount of land you have to use for goats will help you determine how many goats you can raise. To maintain a healthy herd, an acre of land can hold a maximum of seven full size goats or ten mini-breeds. Answers to these questions are the start of your herd management program.

Land/Soil ... If you read chapter one, you know how important it is to survey your land. By knowing your slopes, hills, low-lying areas, and flood prone areas, you can decide what areas are going to be the best areas to build shelters, pen areas, and where to place your feed troughs, waters and more. Determining this now helps to prevent a lot of future changes and redesign.

Normally Dry   After One Day of Rain

If your goats are going to browse at all, knowing your soil and plant composition is key to preparing for the proper nutrition and diet of your goats. Sample soil and natural browse tests should be done prior to getting any goats. Goats are natural browsers and even if they have fresh browse available at all times, they will still need additional hay/feed or supplements. To determine what additional nutrients your goats will need and how much, you need to know what they have available naturally as browse.

To start a good herd management program, have samples of your soil and plants tested for mineral content. Contact your local agriculture agent to arrange for soil and plant testing. Testing usually takes three to four weeks to complete. Once you know what nutrient values you have available naturally for your goats, you can review hay and feed qualities to determine what hay/feed you will use and what minerals may need to be added to the diet.

Fencing … This is the one area you should not trim the budget. Start off with the right fencing now and it will save you money in the long run. Too many times new goat owners go with the cheapest fence they can afford and end up spending more and more money repairing and redoing the fence to get it right.

Not good for goats

Right size holes

Fencing should be sturdy enough not only to keep your goats inside the fenced area, but to protect your area from predators – foxes, coyotes, and even dogs. Even though goat horns may look intimidating, goats cannot protect themselves and are easy prey for most predators.

For smaller breed goats such as Pygmies and Nigerians, a four foot Horse fence with the small holes posted with 4x4's every 8 feet and t-posts at 4 foot intervals will suffice. Goats head butt posts and structures a lot. It is suggested that the 4x4 posts are cemented in the ground. For large breeds, fencing will need to be at least 5 feet high with some type of cross bars at the base of the fence to prevent goats from going under. Pasture fence, barbed wire, field fence, and any fence under 4ft high does not make good fencing for goats; wire cattle panels, horse fence, chain link, wood privacy fence all can work as fencing for goats. Electrical wiring may also be used with the fencing to prevent them from going over the fence. You will be surprised at how high a goat can jump. And make sure if you do use cross bars that they are placed on the outside of the enclosed area. Otherwise your goats will use these as pole vaults to jump over the fence.

Shelters .... No matter what type of area you use, pasture or drylot, goats need shelter. You can build on a budget or build elaborate barns as shelters for your goats. All shelters should have a minimum of three sides and a covered roof to protect your herd from rain and wind.

If you're working on a budget, pallets are a great resource. Using pallets to build a shelter is very inexpensive, easy to build and just the right size if you want to add a lean to ladder out of yet a smaller pallet to allow your goats climbing access. We built our shelters from pallets we found or that were given to us.

Pallets are sturdy, ready-made and reliable. To build a shelter, we placed three pallets together for the sides and one on top for the roof, covered with plywood and roll roofing. Screws should be used instead of nails as nails have a tendency to back out with goats.

Shelters on a budget

The floor of the shelter should be left as dirt. Over the years we have found dirt is so much easier to clean and disinfect. It's natural and free. You can cover the dirt areas with hay for a softer feel and it is easily raked out during cleaning times. You don't want a wooden floor. Goats will pee and poop in their shelters and wood will rot and smell from ammonia build up.

 Another option on a budget includes using extra-large dog igloos as shelters. Igloos are small and can only hold two adult sized goats per igloo. They also must be cleaned out more often than a built shelter, but goats love them because they double as a climber for goats to jump on.

Areas … There are two main fenced pen areas to include in your plans – an area for bucks; an area for does. But if you have any plans of breeding or bringing new goats into an existing herd, you will need additional pen areas. It is best to have more than less and always prepare for expansion of a herd. Remember goats are herd animals, so if you have bucks and does you will have a minimum of four goats, two of each.

Additional pen areas are needed for when does are giving birth, for kids at weaning time, and for quarantining new goats. These additional pen areas may be included as part of your barn design or as individual fenced areas each with their own shelter, feed troughs and waterers. Does giving birth sometimes need assistance with delivery. It is best to prepare for the space you

need now, than find out you needed more at that crucial moment.

A good size pen for one doe is a 4ft x 8ft pen area. Another thing to consider is your weaning age for kids. Goats become fertile around seven weeks of age, but should never be allowed to breed that young. Planning now for separation of kids at this age is very important. Most does have twins or more, a single birth is the anomaly. Your weaning pen should take into consideration the possible number of kids at one time and separation of the buck kids and does once they become fertile. A good sized weaning pen is 16ft x16ft that can be split in half at seven weeks of age.

The quarantine pen area is an integral part of the health and wellness of your herd. A quarantine plan for all new goats for a minimum of two months should be implemented. This time allows you to complete health checkups and monitor the new goat without any possibility of contamination to your existing herd. The quarantine pen is recommended to be at least twenty feet away in all directions from the fence parameters of other pen areas and all barn areas. You also need a quarantine pen for any goats you have that may come down with illness symptoms.

Feed Troughs / Waterers ... The placement and type of feed troughs and waterers you use should be planned based on the setup of your areas, weather, and health considerations of the goats. There will always be a herd queen and a dominant buck. These two will become pushy at feeding time and may need to be fed separately from the rest of the herd to ensure other goats get the proper nutrition.

Feeders should be chin level or higher to prevent goats from pooping and peeing in them. There are many types of feeders available for purchase. Try to choose feeders that are shallow and long over deep and wide.

There are not that many commercial designs for hay feeders. Wire hay feeders tend to allow a lot of wasted hay. Many goat owners have found making their own homemade hay feeders better than a commercially bought hay feeder. A simple one that we have found to work very well is a large plastic tote tub. These can be screwed to a wall above goat head level with several 10 inch circular holes cut out on sides and bottom of the tub. Hay stays in place and goats can easily reach up to pull out hay as they eat. Just make sure the cuts are sanded smooth to prevent cuts and scratches.

Waterers can be simple buckets, automatic water tubs, or elaborate tube flow systems. Buckets will have to be emptied and refilled at least every two days to prevent algae buildup. Automatic waterers are a great investment but must have a water line accessible to the tub and may need repair at times. Tub/nipple flow systems are more expensive but a good investment if you are going to have new kids every year. Kids easily adjust to the tube system as it simulates the natural milking position.

Whatever you decide for your barnyard basics, you must also answer the question of your need for electrical and water access in your areas. Barns encompass such a large area it is most often essential to have ready access to both. But even electricity for lights and water access can be beneficial in smaller pen areas, especially in your birthing pens and quarantine pen. You never know when your assistance is going to be needed in the middle of the night. Plan now for where you need electric or water access so you don't have to redesign later.

Goat Management

Now that you have the barnyard basics down pat, it is time to make a goat management plan as part of your Herd

Management Program. Goat management covers identification, disbudding, castration, regular checkups and/or vaccinations, routine maintenance, and basic record keeping.

   Identification … If you show goats or have registered goats, they must be tattooed for identification. But even if you decide you do not want to show or own registered animals, identification is important for your herd.   Tattooing your animals not only helps identify which animal is which, but allows for easy identification if the animal is ever lost or stolen. It also allows you to maintain accurate records for breeding and lineage.

   For registered and show goats you must join a registry for your breed type. You will need to register your herd name and acquire a herd prefix for tattooing. All of your registered goats will use the herd prefix as part of your tattoo system. If you do not register goats, you can choose any tattoo lettering or numbering system of your choice.

   The right ear or right half of tail web is used for tattooing your herd prefix, or if unregistered this is where to tattoo your own letter/number system. The left ear or left tail web is used for birth information. The left versus right is known when you are facing the same forward direction as the goat, your left is the left and right is right. The general consensus for left tattooing includes a letter for the year of birth. Letters G, I, O, Q and U are not used. Starting in 2013 the birth year letter is D and continuing alphabetically to Z in the year 2030. After the birth year letter a number is used to identify the birth order. As an example, a baby goat was the second birth in the year 2015, the left ear tattoo would then be F2 – F for 2015 and 2 for second kid born.

Tattooing is simple and there are several tattoo devices available. You can use a hand powered crimper style or an electric pen device and price can range from $50 to $150 on average. My personal preference is an electric tattoo pen tattooing the tail web area. The tail web tattoos seem easier on the goat and are easier to read. The only disadvantage to tattooing the tail web is using the crimper style tattoo devices. You would need a very small kit in order to tattoo kids.

Disbudding … Is a must for show goats, but also adds the benefit of protection for you and your herd. Goats will head butt almost everything and at times can use their horns against other goats and even humans. Not all goats will use their horns in this manner, and our preference is to cull or sell those that do. But some may say 'why take the chance'. Decide whether or not you want goats with or without horns prior to purchasing any goats.

Disbudding is the removal of the horn bud before it ever starts to grow. It is done when the goat is a few days old. Dehorning is removal of the horn after it has grown and is usually not recommended. There is a major blood vessel which runs up through the horn and removal of the horn usually requires surgery by a veterinarian. Polled is another term you will hear often. It is a goat who is born naturally without horns and no horn development. Many goat owners choose to have polled goats for show to alleviate the need for disbudding.

Castration … Any goats not kept as a herd sire or that are not going to be sold as an intact buck should be castrated. Bucks have a natural smell that many owners find intolerable. Making a buck a wether (castrated male) eliminates most of the bucky smell and also allows to keep a goat in a situation where an intact male is not wanted. But it also increases the chance of urinary stones to form, and a castrated goat has special dietary needs to maintain a healthy urethral process.

Castration is a very simple process and can be done by the goat owner or by a veterinarian. It should usually be done around seven to eight weeks of age to allow for full urethra development, although many owners castrate at just a few weeks old without further issue. There are many methods of castration and the option you choose to use should be planned for from the beginning.

Normal Goat Health

•   Goats Life span: Does live 12 years on average but goats can live over 18 years; Wethers (non functioning buck) live 11-15 years on average age; Bucks live 8-10 average age.

•   Goats Temperature – normally range between 102.5 and 104;

•   Goats Respiration – breaths average 12 to 25 per min;

•   Goats Pulse rate is 60 – 80 beats per min;

•   Goats Stomach movements is 1 – 1.5 per min;

•   Goats reach sexual maturity by 7 weeks to 8 months depending on breed;

•   Goats Fertility Cycle is 18 to 22 days, receptive to buck for only 12-36 hours, ovulation occurs the last hour of standing heat;

•   Goats Pregnancy Duration lasts 148 to 156 days;

Determining a Goat's Age

If you do not know a goat's age, you estimate its age by teeth. To estimate (not a guarantee of age) a goat's age, look at the bottom front teeth.

- On kid (under a year old) the lower front teeth will be small and sharp.

- On a yearling goat (1yr-2yrs) their two middle lower front teeth are replaced by larger adult teeth around 12 months of age.

- At two to three years old, they lose the two teeth next to the middle lower.

- By the time a goat is four years old, it will have six adult teeth on the front lower bottom with only two baby teeth left.

- All eight lower front goat's teeth will be there by age five.

- Goat's teeth will spread as a goat ages over five, and will generally start losing teeth after age ten.

Health Checkups … No matter what the size of your herd is or what it will be, every goat needs regular checkups and possibly vaccinations. Plan to do at least monthly checkups on your goats. At least two health checkups should be performed on any goat while in quarantine. To start, you should know a little about normal goat health and the age of your goats.

Your monthly health checkups should be performed on a regular schedule. Here's an example of what we do – we chose the first Saturday of the each month. First the goats -

We wear gloves throughout the examination and complete

one full exam on one goat, change gloves and then go to check another goat. You should check:

• Overall appearance for walk, gait, and healthy appearance;

• Eyes for clear look and proper under lid color;

• Ears for mites, infection, cuts and scrapes;

• Mouth, gums and teeth check for sores, swellings and color; Teeth for proper formation, abscesses or missing teeth –

we brush their teeth with non-fluoride toothpaste using a different toothbrush for each goat.

- Fur for thickness, shine, and use a brush against the lay pattern to check for fleas, mites or ticks;

- Body - we run our hands across every part of their body and especially around the neck and other lymph node areas for swellings, cuts, abrasions, scratches.

- Feet - we complete a full hoof trimming and inspection for any sign of hoof rot, cuts, or lodgings.

- Fecal - we take each goat's temperature, then we gather fecal material from all goats and complete a fecal exam for worms.

- NOTE: Goat's Temp – we use a digital thermometer. Keep this thermometer for goat use only – DO NOT use on humans. Always clean thermometer by dipping in bleach water and drying off before using on another goat.

We also use a Body Condition Score (BCS) to rate the overall condition of the goat. To determine a BCS while feeling the goats body you should focus on the rib area, the tail head, the loins and backbone. Run your fingers along these areas pressing down with three fingers. You are checking for the amount of fat buildup in these areas. Practice makes perfect and the best way to learn is by checking goats that you know look overweight or too thin first.

Our BCS scoring ranges 1-10. In goats a score of 8-10 is rarely seen.

BCS	What it Means
1	Near death, a goat with a one is most likely not able to stand on its own.
2	Extremely thin and weak, a goat with a two is severely ill
3	Thin but not weak, ribs are visible, a goat with a three needs medical attention and further checkups
4	Slightly thin with some ribs visible, a goat with a four needs additional nutrients and dietary review
5	Healthy with some fat, a goat with a five is healthy but may need additional nutrients
6	Healthy looking, no ribs visible, moderate amount of fat
7	Slightly overweight, no ribs visible but more pressure is needed to feel, needs dietary review
8	Overweight, fat is noticeable, a goat with an eight needs to be place on a diet
9	Obese with a lot of fat, a goat with a nine has little movement and looks fat, needs further checkups and strict diet
10	Severely Obese, a goat with a ten has trouble getting up and moving, needs medical attention

Once done with the above checkups on all goats, this is the time to dispense any medications needed.

Vaccinations … Some goat owners believe regular vaccination helps prevent disease in a herd. This is a choice you will have to make before getting goats. We feel that with regular health checkups and a proper Herd Management Program, illnesses should be treated on an as needed basis to allow a goat's natural immunity and antibodies to build up a healthy system. However, because vaccinations can be valuable for some goat owners, they will be included in the Disease and Illness chapter.

Routine Maintenance … Along with your monthly health checkups, it is a good idea to complete routine maintenance and inspection on your goat facilities. Proper sanitation is not only healthy for your goats but can prevent many diseases and illnesses.

Non-Goat Inspections – Check and repair

•   Inspect all shelters for wear and tear – broken boards, rotting boards, leaks.

•   Inspect all feeders and stalls for wear and tear – broken boards, rotting boards, leaks, and ensure feeders are at correct height.

• Inspect all waterers for proper function, remove any leaves from water source, replace water and add ACV to bucks water supply.

• Inspect hay stacks for looseness and mold. If wooden, check for broken or rotting boards.

• Inspect mineral blocks and replace as needed.

• Inspect all fences for broken sections or those in need of repair.

• Remove and replace all bedding in all areas and rake out all enclosed pen areas and areas around shelters. We put the hay and raked material into a pile to make homemade fertilizer for plants. Allow it to ferment for six months, tossing monthly, and then use bottom of pile to place around plants.

If you have goats with certain diseases as discussed in the Disease and Illness chapter, there may be times that all materials will have to be sanitized or even burned to prevent passing the infection to other goats in your herd. Be prepared to have a plan in place to deal with this if it should ever occur.

Record-keeping ... not only helps you manage your herd better, but will also help you see in writing what your expenses are for keeping goats. There are several programs available to help you maintain records, or you can keep it simple by using a spreadsheet program or writing tablets. Whatever way you go, you need to keep accurate records.

The best way to give you an idea of what information to keep is to provide you with an example. We use Excel spreadsheets to maintain our records. In Excel we have a tab for listing all of our goats by name with their breed type, name, identification tattoo, year of birth, weight at birth if born on our property, and sex – buck, wether, or doe.

You can use the spreadsheet tabs to monitor heat cycles and your breeding programs. On these tabs we use only the identification tattoo and then have columns for dates of heat cycle, and projected heat cycle dates; or date of breeding and with what buck and projected delivery date.

Monitoring sick goats and goats in quarantine is easy using tabs. We have a tab for medical treatments and quarantines which includes columns for suspected illness, two checkmark columns for healthy checkups, fecal test date, checkmark column for worms, and treatments given.

Another great reason to keep records is to track your expenditures for startup and monthly expenses. You can track how much you spend on feed, hay, supplements, medicines, just about everything. Use it like you would a household budget to help you eliminate wasteful spending, and to determine your actual costs versus income ratio if selling goats. Most goat owners quickly realize that expenses quickly outweigh any income.

If you show goats, you can create a spreadsheet for upcoming shows and record awards and notes given from the judges. Also, a spreadsheet program allows you to create a one page checklist for each goat with a note section to use for your monthly checkups and routine maintenance. They can be printed and taken with you to mark during your checkups and will give you a visual reminder of areas of concern and possible treatment needs. After recording your results, you could file them in folders for later retrieval or review. You will find many reasons why you need to maintain records, and this is a very important part of your herd management program.

Chapter Three - Nutrition and Feeding

Goats are ruminants and browsers. Their dietary needs are different than other animals, even different than other ruminants. Bucks, wethers, pregnant does, and lactating does all have special dietary needs. Many diseases and illnesses in goats are brought on due to improper feeding or from feed made for other species. Proper nutrition and feeding is very important.

TDN or Total Digestible Nutrients (which is a measure of energy and quality of feed, the Protein content, Calcium content, Phosphorus content, and daily amount of feed based on % of body weight) are important to review. The information which follows are the minimum requirements for healthy goats. Young goats a year old and under should have a TDN between 65-68%, Protein at 12-14%, Calcium .4-.6%, and Phosphorus .2-.3%, DF 3.2-3.7%; for bucks TDN should be least 60%, Protein 11-12%, Calcium .4-.6%, Phosphorus .2%, DF 2-3%; for pregnant does without milk TDN of 60%, Protein 10-11%, Calcium .4% and Phosphorus .2%, DF 2.4-3%; and does in milk TDN of 65%, Protein 14%, Calcium .6%, Phosphorus at .3%, DF 2.8-4.6%.

Bucks and wethers can develop urinary calculi (UC) if overfed grain or alfalfa hay and not protected. Prevention of UC is handled by following strict calcium to phosphorus ratio. The recommended calcium to phosphorus ratio is 2 to 1. However, we have found 3 to 1 to be better. We also add Apple Cider

Vinegar to the water for bucks and wethers and have found it to be a great preventative.

Other minerals are in general based on a goat's daily diet. These minerals include Sodium, Potassium, Chloride, Sulfur, Magnesium, Iron, Copper, Cobalt, Zinc, Manganese, Selenium, Molybdenum, and Iodine. Although low levels of any mineral or vitamin can cause issues in a goat, copper and selenium issues are becoming more recognized. Copper and Selenium are very important to a goat's health, and are naturally deficient in most areas and in feedstuff so are therefore lacking in a goat's diet.

The B vitamins are also very important in maintaining the health of your goat. B1 and B12 should be given to any goat that is sick, acts a little off, stops eating, or whenever there is a change in a goat's diet. B1 Thiamin provides energy by converting blood sugar into energy. It keeps mucous membranes healthy and is essential for the nervous system, cardiovascular and muscular function. A lack of thiamin in the goat's system can be devastating and may become fatal. Goats produce their own thiamin but when there is a thiamin deficiency, the result can be swelling in the brain, temporary blindness, lack of coordination and staggering which can result in death. Treatment with B1 can show drastic improvement within a few hours.

B12 cobalamins keep nerves and red blood cells healthy. It is responsible for the smooth functioning of several critical body processes. It converts carbohydrates into glucose leading to energy production and a decrease in fatigue and lethargy. B12 helps in healthy regulation of the nervous system, reducing depression, stress, and brain shrinkage, and it helps maintain a healthy digestive system. It is essential for healthy skin, hair, and hooves. It also helps in cell reproduction and constant renewal of the skin.

When a goat stops eating or when the digestive system is not able to absorb this vitamin well, it is usually from bacteria growth in the small intestine, or a parasite, a deficiency in vitamin B12 occurs. B12 deficiency can be fatal in goats and long-term treatment may be necessary. B12 deficiency results in illnesses like anemia, fatigue, weakness, constipation, loss of appetite, weight loss, depression, poor memory, soreness of the mouth, vision problems, and a low sperm count.

Hay should be the mainstay of your feed program after natural browse. It should not be measured but kept out fresh at all times. There are some hays that should be avoided if possible and not used as part of a goat's diet. These include Fescue, Johnson's, and Sorghum hay types. If you don't already know the nutritional value of your hay, your local agricultural extension office can assist you in getting it tested. Quality of hay and nutritional value can vary among producer.

The table on the next page is a general guide for different hay contents.

	Crude Protein %	Calcium %	Phosph. %	Magnes. %	Ca:P Ratio
Alfalfa	16	1.28	0.24	0.3	5.3:1
Barley	7.8	0.21	0.25	0.12	0.8:1
Bermuda grass	9.2	0.43	0.16	0.16	2.7:1
Kentucky Bluegrass	9.1	0.4	0.27	0.19	1.5:1
Bromegrass	8.7	0.32	0.15	0.09	2.1:1
Red Clover	13	1.22	0.22	0.34	5.5:1
Grass	8	0.44	0.18	0.2	2.4:1
Lespedeza	13.8	0.78	0.25	0.2	3.1:1
Oats	8.6	0.29	0.23	0.26	1.3:1
Orchardgrass	9.4	0.34	0.23	0.16	1.5:1
Prairie Grass	7.9	0.51	0.17	0.22	3.0:1
Rye	7.9	0.31	0.18	0.12	1.7:1
Timothy	6.8	0.38	0.17	0.11	2.2:1
Wheat	7.7	0.13	0.18	0.11	0.7:1

Although goats may seem to be picky, if the hay is part of their main diet they will eat it. Because goats are browsers, they will eat more of the hay if you can hang it above or just at head level. Goats tend to waste a lot of hay, so choosing a proper hay rack is important. We have found in herds under twenty goats, purchasing square bales is more economical than purchasing the large round bales and there is much less waste.

After determining what hay you have available in your area and which hay you will use, you may still need to offer other feed and/or mineral supplements to ensure proper nutrition based on what you have available. The table below provides general nutrient values for various feeds.

	% TDN	% CP	% Ca	% P
Alfalfa meal	69%	28%	2.88%	0.34%
Alfalfa pellets	61%	19%	1.42%	0.25%
Barley grain	84%	12%	0.06%	0.38%
Beet pulp, dry	75%	11%	0.65%	0.08%
Corn grain	88%	9%	0.02%	0.30%
Corn silage	72%	8%	0.28%	0.23%
Cottonseed meal	85%	46%	0.13%	0.55%
Cottonseed, whole	95%	23%	0.17%	0.68%
Fish meal	74%	66%	5.50%	3.15%
Ground ear corn	82%	9%	0.06%	0.28%
Kelp, dried	32%	7%	2.72%	0.31%
Limestone (ground)	0%	0%	34.00%	0.02%
Molasses	75%	6%	0.97%	0.10%
Oat grain	76%	13%	0.05%	0.41%
Soybean hulls	77%	13%	0.55%	0.17%
Soybean meal	84%	49%	0.38%	0.71%
Soybeans, whole	93%	40%	0.27%	0.64%
Trace mineral salt	0%	0%	20.00%	8.00%

Everything needs to work together to provide the proper nutrition for your goats. Do not go by recommendations for hay, feed, or supplements that someone tells you to use for your goats. There is no diet, feed or hay, that anyone can recommend that will be perfect for your herd because what you have naturally available is not the same as what they have available. This is where land, soil and browse/hay testing is important. To save overall costs on feed and to avoid purchasing several different feeds for each goat's need, it is easier to choose a hay type that covers the basic nutritional needs of your goats as a herd, then add in a complimentary feed and/or supplements for those goats that need additional nutrient amounts.

Also, what may be okay for horses, cows, sheep, or pigs will not always be safe for goats to eat. Goats have a natural instinct for what they can or cannot eat. If you watch your herd long enough, you will see your Queen Goat choose where and what the herd should eat. With that said, NEVER willingly allow your goats to eat something that is known to be poisonous them. If possible, it is best to remove any poisonous plant from the area where your goats will wander. The table on the next page is not all-inclusive and there may be poisonous plants in your area which are not listed.

Poisonous plants to goats

African Rue	Cassava	Fiddle-neck
Andromeda	China Berry Trees	Flixweed
Avocado	Choke Cherry	Fuschia
Azalea	Datura	Hemlock
Brouwer's Beauty	Dog Hobble	Holly Trees/Bushes
Boxwood	Dumb Cane	Ilysanthes
Burning Bush berries	Euonymus berries	Japanese pieris
Calotropis	False Tansy	Japanese Yew and Yew
Lantana	Lupine seeds	Oleander
Larkspur	Madreselva	Pieris Japonica
Lasiandra	Maya-Maya	Red Maples
Lilacs	Monkhood	Rhododendron
Lily of the Valley	Milkweed	Rhubarb leaves
Mountain Laurel	Coriaria arborea	Wild Cherry

Chapter Four - Breeding

With breeding you can just let nature occur, or you can prepare a breeding program. This chapter discusses using a breeding program. If you are using the suggestions we have offered and have already begun monthly checkups as part of your herd management program, you will also be able to incorporate a breeding program rather easily. A breeding program helps you improve your herd, allows you to determine what goats you want to breed, lets you choose when you want to breed, and improves the health of your goats for breeding.

No matter what your original reason for raising goats, when breeding, the primary goal should be to improve herd quality even if your intention is to sell the kids. Working to produce better qualities in your offspring will not only allow you to offer clientele quality goats (which could be sold for a higher price), but if you choose to keep some of the offspring you will improve your own herd line.

Always look for a buck that has characteristics that can improve upon your doe's faults or areas which need improvement; same with choosing a doe to breed to a buck.

They should complement each other in areas of conformation: where one lacks the other excels. Get to know your breed standard and what you're looking for to improve upon even if you are not showing goats. Producing a goat with better breed qualities helps everyone.

Use the body condition scoring system discussed in the Herd Management chapter to determine the overall condition of a goat prior to breeding. Scores range from 1 to 10 with 1-3 being too thin for breeding, 4-6 being just right for breeding, and 7-10 being too fat to breed. Overly thin or fat goats are not used for breeding because both of these characteristics generally cause health issues for the goats during breeding, pregnancy and afterwards. In bucks, body condition has an effect not only on its health but its desire to breed as well. Although the system may seem complicated, it is rather easy to use.

Depending on your breed of goat, the breeding season can run from August to January or throughout the year like with mini-breeds of African Pygmies or Nigerian Dwarfs. Some breeds that normally cycle seasonally can also breed out of season. A doe in season will go into heat on average every 18-22 days, but can be forced into heat by the introduction of a buck, and increase in available daylight by using lamps. It has been shown that does closer to the equator change from seasonal breeding to year-round cycles. A doe will normally go into heat between 7-10 days after being introduced to a buck if not currently in heat. Does in heat are usually receptive to the buck for 24-36 hours with ovulation occurring in the last hour, or 12-36 hours after onset of heat.

To know when your doe is in heat you can monitor her cycles and record heat cycle dates, or you can monitor your doe's behavior. Most does will have a noticeable change in behavior during heat. They will either become friendlier or more aggressive. Some does will start to mount other does or even the bucks. They will arch their back, flicker their tails, and rub up against other goats. Does in heat will usually have a clear to white discharge and you will notice her tail end being wetter than normal. However, some does show no signs of heat. For does that show no signs of heat, you can take their temperature daily. When their temp rises a degree or two, they are going into heat.

If you have your own herd sire and doe you want to breed, you can start to prepare them both for the act of breeding. They should be in the best possible health prior to breeding. Does should be given additional nutrients to help with ovulation and maintaining the fetus throughout pregnancy. Bucks also need additional nutrients to ensure healthy sperm for fertilization even if you choose to use artificial insemination. There are some old wives tales that say how or what you choose to feed your buck in the few weeks prior to breeding can increase the likelihood of producing does. Although they are fun to review and maybe even try, there has been no scientific research to prove the suggestion true.

With your own herd sire, breeding is as simple as bringing the buck to the doe and leaving them together for a few days during heat. If you have a controlled breeding program you can watch to ensure fertilization has occurred. To know when the deed is done, watch for at least two good consummates. A doe that accepts a buck while in heat will arch her back afterwards and then she will squat as to pee, sometimes thrusting forward without actually peeing. The more you see the act, the easier it will be to recognize when fertilization has occurred.

If you do not own a buck, you can breed your does by artificial insemination or by using a stud goat. Stud fees vary by area but in general are in the range of $100.00. You save money by not having to feed or house an extra goat at home. With stud services, you must take into consideration the ranch conditions of the buck's home. Never take your doe to a buck's home without first being able to inspect their facilities. You don't want to take a chance with your goat's health or your herds' health when you bring your doe home. Even if you are able to bring the buck to your home, you will still need to inspect where the buck came from to ensure the health of your goats. The buck's owners should be asking you questions as well. If they don't I would be wondering why they weren't.

Questions to consider – Do they vaccinate their herd and what for? Have they had testing on their herd for CAE, CL and other diseases? What were the results? Have they had any known health issues at any time on their property – such as sore mouth? Do they have references from others they have studded for and can you contact them? Do they have a specific area set up for breeding and is it away from other goats? How long do they allow the goat to remain on property? Do they offer a guarantee for re-stud if the first attempt does not succeed?

If you are using a registered goat and plan to register the kids you should also ask questions about registration. Will you maintain breeding rights of the kids produced? Will they sign the paperwork for registering the kids? Are they going to want to keep a kid from the breeding? There have been cases where one thing was agreed upon and then an owner changed their mind and wouldn't complete the deal. Be smart and be proactive – get everything in writing prior to the act.

Artificial insemination (AI) is becoming more useful for impregnating does when you don't own a buck. You don't have to worry about transporting your goat somewhere and you don't need to bring a buck to your property. You control the act of fertilization and can ensure it is done at the proper time. You also may be able to purchase semen from a quality buck that you would otherwise not be able to use. But you still need to know about the health of the buck the semen comes from. It is best to ask the same questions as you would if using a stud service. Artificial Insemination may be more costly than using a stud service to begin with but the benefits of using AI costs less in the long run. Startup costs can range from $300 to $1000 just for the needed supplies prior to semen purchase. You may also want to purchase hormone therapy to synchronize your does' heat cycles if you are trying to impregnate more than one doe. The use of Prostaglandin and Progesterone can help synchronize does. Disadvantages of AI include having to regulate/monitor heat cycles very closely and the fact that *semen is only viable for a few hours.*

Take into consideration all the reasons, including costs, prior to breeding. You must decide which avenue is best for your herd whether you own a buck, use a stud service, or use artificial insemination. There are pros and cons which need to be considered for each. Any option you choose will be the right one for you, and if not you can change it later.

Chapter Five – Pregnancy and Kidding

Pregnancy lasts 148 to 156 days. If you have used a breeding program you will be able to track your doe's pregnancy and due date. Your doe should not be bred until at least two months after giving birth and should be dried up at least ten days prior to breeding. Worming, if needed, should be completed prior to breeding or after 75 days pregnant with a wormer safe for pregnant does. Once your doe has become impregnated, she will no longer cycle into heat. She may have a change in attitude, becoming more protective or aggressive.

		148 day Breeding Calendar			
Bred Day	**Kid Day**	**Bred Day**	**Kid Day**	**Bred Day**	**Kid Day**
1-Jan	29-May	30-Apr	25-Sep	5-Sep	31-Jan
9-Jan	6-Jun	8-May	3-Oct	13-Sep	8-Feb
17-Jan	14-Jun	16-May	11-Oct	21-Sep	16-Feb
25-Jan	22-Jun	24-May	19-Oct	29-Sep	24-Feb
2-Feb	30-Jun	1-Jun	27-Oct	7-Oct	4-Mar
10-Feb	8-Jul	9-Jun	4-Nov	15-Oct	12-Mar
18-Feb	16-Jul	17-Jun	12-Nov	23-Oct	20-Mar
26-Feb	24-Jul	25-Jun	20-Nov	31-Oct	28-Mar
5-Mar	31-Jul	3-Jul	28-Nov	8-Nov	5-Apr
13-Mar	8-Aug	11-Jul	6-Dec	16-Nov	13-Apr
21-Mar	16-Aug	19-Jul	14-Dec	24-Nov	21-Apr
29-Mar	24-Aug	27-Jul	22-Dec	2-Dec	29-Apr
6-Apr	1-Sep	4-Aug	30-Dec	10-Dec	7-May
14-Apr	9-Sep	12-Aug	7-Jan	18-Dec	15-May
22-Apr	17-Sep	20-Aug	15-Jan	26-Dec	23-May
		28-Aug	23-Jan		

A pregnant doe requires additional nutrients to accommodate the growing fetus. She should have free access to hay and minerals, but should not be overfed feed or grain. Pregnant does that gain too much weight or not enough weight during pregnancy can develop health problems. Pregnancy disorders such as ketosis and toxemia are discussed in the Disease and Illness chapter. If the does are healthy prior to pregnancy, we do not change their diet during the first four months of pregnancy with the exception of offering high quality alfalfa hay.

However, the last six weeks of pregnancy is when most of the development occurs and our does are given an increase in pellet ration from ½ cup to 1 cup per day or half a pound to one pound per day. We also add probios or a natural yeast to help aid in digestion during this time, and additional minerals offered free-choice. A healthy pregnant doe should maintain a healthy Body Condition Score throughout pregnancy, and should gain between 15 and 40 pounds during the last six weeks of pregnancy, depending on the number of kids she is carrying. The dietary increase for your pregnant does should be based upon your current nutrition plan.

At five weeks prior to due date selenium and copper supplements should be given if you are in a selenium or copper deficient area. You will notice more changes in your doe during the last month of pregnancy. She will start stretching more as the kid or kids move into position; she may moan or make noises towards her belly. There may also be some clear to whitish discharge off and on during this time; this does not mean delivery is at hand. If you notice any bloody discharge, consult a veterinarian as soon as possible.

At four weeks prior to delivery date a CD&T and Pneumonia vaccine should be given if you vaccinate your herd, and you should do a full hoof trim. At three weeks prior to due date you should start dosing your doe with Vitamin E daily, and at two weeks prior make sure you have your birthing kit prepared and ready.

Suggested birthing kit supplies (other emergency supplies may be needed):

- Two rolls of paper towels;
- Eight puppy training pads;
- Two large outdoor garbage bags;
- A flashlight with extra batteries;
- Scissors;
- Plastic disposable gloves;
- A bottle of Betadine or Iodine;
- Vaseline or other lubricant;
- Small bowl or empty film canister;
- Large bucket of water;
- String or dental floss;
- One jar of molasses and water;
- Two large cloth towels;
- A baby bottle and nipple;

- Coffee;
- Store bought milk or goat colostrums;
- At least four kid sweaters;
- Eight hand warmer packets or a blow drier;
- A scale to weigh kids.

If you have separate areas as suggested in the Herd Management chapter, the birthing pen should be cleaned and have fresh bedding placed down a week prior to due date. The pregnant doe should be moved into the birthing pen after it has been made ready. Because goats do not like to be alone, another doe can be penned next to the pregnant doe to lower the stress level if you have more than one stall area.

At this time, if you choose to do so, you can shave the doe's udder, tail, belly, and vaginal area. Shaving these areas makes it easier to check ligaments, clean up after birthing, and helps kids find the teats. You should begin checking the tail ligaments daily to better determine kidding time. The ligaments feel like hard pencils in an upside down V shape starting about two

inches up from the tail head running downward on the sides of the spine towards the pin pones.

To check the tail ligaments run your hand about one inch from spine with thumb on one side, fingers on the other, downward towards pin bones.   As delivery gets closer, the ligaments will soften and sink in and the tail head will seem to rise. Once you

no longer feel the ligaments, the area looks sunken, and your fingers and thumb can easily touch around the tail head, labor usually occurs within the next 4 to 12 hours.   Make sure your fingernails are trimmed prior to delivery in case your assistance is needed.

   Kidding is a stressful time not only for your doe but also for you. It is a nerve-racking experience even if you have gone through it many times before. If this is your first kidding season, it may be helpful to have a mentor goat owner to help prepare you for the birthing experience. Although the books and information you review can help you understand what to expect, the reality is the best learning experience comes from witnessing the event itself. So if you get the chance to witness a normal kidding prior to your own goat's delivery, take it.

There are many goat owners that allow their does to deliver without them present; and then there are owners who want to be there at every birth. In most deliveries everything runs smoothly without any issues, but you always need to be prepared and ready for when things go wrong. There are many 'normal' variations of delivery because there really is no one exact scenario of a 'normal' delivery. So, in this chapter we'll discuss a normal delivery and some common issues that may occur which require your assistance.

Typical signs of early labor include the doe becoming more restless, eyes seem distant or staring off into space, increased pawing at ground or bedding, crying or talking to her side. The doe lays down and gets up more often, udder has filled out and is glossy and tight, a stringy egg-white looking discharge or amber goo, vaginal opening looks more pronounced and elongated, or there is a change in attitude. Any one, all or none of the above can be symptoms of a doe about to deliver.

A doe can give birth in different positions; no doe is the same. Some may stand and deliver, while others may lie down.

What you must know are the normal kidding positions, and what to do if assistance is needed. Always have your birthing kit and emergency supplies at hand, and it is always best to have the telephone number of a good goat veterinarian for when needed.

The first thing you may notice is a bubble or water sac coming out of your doe. The bubble may appear, disappear, and reappear; it may break before showing, break during kidding, or break afterwards. If you never see a bubble, don't worry, it most likely broke inside the doe and this is normal too. Once a doe goes into active labor you should see noticeable progression within 30 minutes. If there is no progression in this time or delivery within one hour from start of active labor, the doe will need your assistance.

With or without the water sac, you should look for a kid in normal delivery position. The normal position is the two front hooves pointing down first, then nose and head on top.

There are two normal positions when dealing with singles or multiples. Just remember front feet – hoof soles pointed downward, back feet – hoof soles pointed upwards.

Normal Single Positioning

Normal Twin Positioning

Common kidding positions that need adjustment are when one leg or two are bent back inwards, head is bent backwards, a breech position with butt first, an upside down position, or when multiple kids are facing the same direction. All of these positions need a manual adjustment for you to get them into one of the above listed positions. To assist in these positions, you will need to make sure your own nails are trimmed. Then cleanse your hands and arms with betadine or iodine. Cleanse the doe's vaginal area. Glove up, ensuring gloves are sanitized as well, and lubricate gloves and the doe. Picture the correct position in your mind and think about what you are feeling. Remember to work with the kid <u>as the mom is contracting, not when she is relaxed</u>. This will help prevent uterine tears.

For a kid who is in the normal position but with one leg bent backwards, gently reach in and correct the position of the one leg, then pull alternately on legs until in correct position. If both legs are bent backwards, with your hand follow the kid's head to chest and armpit. Use a finger to hook one leg at the bend and maneuver it forward. Repeat for the other leg, and gently pull alternately on legs until in correct position. Remember to pull out and downwards. All of this sounds complicated, but it is not; just remain calm.

When a kid is in a position with its head bent backwards, push the kid gently back in and maneuver the head into the correct delivery position by placing your hand on the top of the head and lifting up and forward. Never try to pull a kid out with its head bent backwards. If the doe is not cooperating, a veterinarian may be needed. It may also help to have someone lift the doe's back legs to make it easier to push the kid back in. The use of a lamb puller can help keep the kid's head in the proper position during delivery.

Sometimes a doe can deliver a kid who is in a breech position without assistance, but in most cases it is better to rearrange the kid into the proper position. Reach in and locate the hocks or hooves under the tail. Gently push the kid back in more and maneuver each leg until it is in correct position.

Upside down kids or multiples that are tangled require more assistance and at times a veterinarian should be called. If you have the confidence to rearrange these kids, or a vet is not available, you will need to mentally picture what you are feeling. Upside down kids will need to be flipped gently and then maneuvered into correct diving position. For kids that are intertwined or facing the same direction coming at the same time, you will need to really connect what you feel with how they each are positioned. Rearrange their positions as they should be and maneuver them to allow one kid to be delivered before another.

If you notice an amber or dark yellow colored water sac at delivery this usually indicates assistance will be needed during delivery. The coloring is due to the kid excreting its first bowels while being delivered and is a sign of stress. Be prepared to help get the kid out quickly. Do not confuse the water sac with the afterbirth. Afterbirth may normally contain amniotic fluid which could also be amber colored or dark yellow.

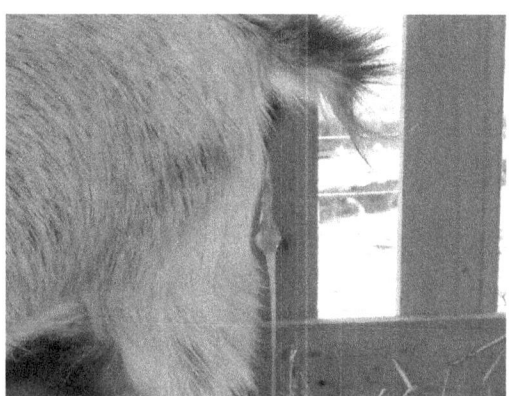

Once a kid is delivered the doe will normally begin to clean it off. If she is being attentive to the kid it is best to allow her to continue without interruption. This helps create a bond between the doe and kid. You may intervene either after she has cleaned the kid, or when she begins to deliver another kid.

Once all kids have been born, the afterbirth should begin to expel. It can take up to 12 hours for all placenta remnants to pass. You may notice another water sac filled with liquid. This is also normal and not a sign of concern. Some does will eat the afterbirth and you may never see it, others will leave it alone.

Here are some good birthing pictures:

Doe cleaning new kid   **Afterbirth Sac**

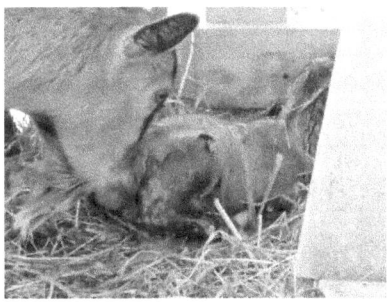   

Checking on new kid   **Natural Bonding**

Contraction   Nursing and Cleaning

Although mom usually takes care of kids after delivery, you can assist her by helping to dry off the kid with a towel. But do not dry off the kid completely until you have allowed mom to have some bonding time. Check to make sure the doe tore the umbilical cord short enough – no more than two inches in length. If it is longer, go ahead and cut the cord to the right length. Next dip the cord into your betadine or iodine solution. If you have had to cut the cord or mom is an aggressive licker, you should tie the cord off with the dental floss or string to make sure it doesn't get infected.

Sometimes there are issues with the umbilical cord of a newborn kid. At times the cord itself may become extended, engorged, or filled with blood – or worse it may be an umbilical hernia and not an issue of extended cord growth.

For extended, engorged or blood filled cords – These will normally heal on their own but if the cord is causing discomfort or is in the way of normal movement for the kid, the cord can be clamped up. Make sure to re-dip the cord after clamping and give 1cc of penicillan to prevent infection.

When to worry and take the kid to a vet – If the cord is bleeding out, black, smelly or shows signs of hernia or gangrene, take the kid to a vet for observation and possible need of hernia suturing. If in doubt, always consult a veterinarian. Umbilical hernias can leave an opening where the intestines can come out. It is not something you want to take a chance with if unsure and can be easily corrected.

If it has been a difficult birth or the kid seems weak, check the kid's gums. If they are a dark red or purple color this means the kid was oxygen starved during delivery. Take a towel and rub the kid's face and nose, then pinch its ear or tail to get the kid to scream. This helps the kid get the needed oxygen. You can also use a nose syringe or piece of hay up the nose to get the kid to sneeze. If it is cold outside, you can put a sweater on the kid to help keep it warm. A cold kid will also seem weak or lethargic. You can dose a newborn with Selenium, B1 and a Tetanus antitoxin to help aid against illnesses.

Once the kid seems alert and active, place it on a puppy pad in front of mom to allow her to bond and clean the kid some more. Watch her carefully. Some does, especially first timers, need your help to bond with the new kid. You can help by not only placing the kid in front of her, but also by taking some of the birthing fluid from the kid and wiping it on the doe's nose and then putting baby up to her nose for her to see the kid. This should get her mothering nature to kick in and she should start cleaning the kid.

Bonding is very important for the doe and kid. But sometimes you can do everything right and mom still will not bond. A doe that ignores the kid, runs from the kid, or head butts the kid is a mom doesn't know how to bond or doesn't want to bond. You will need to step in and take over caring for the kid in this case. Some does will paw at their kids. This is not a sign of mom not wanting to bond. In fact it is a way mom is trying to bond by getting her kid up and moving and trying to get the kid to nurse.

A kid needs to get colostrums within an hour from birth. Check the doe's udder to make sure milk is available. Sometimes the teats will have a plug in them that will need to be removed so the kid can nurse. Gently squeeze each teat until there is a good flow of milk. Most often the baby will find the teat on its own and begin nursing. Does will lick the rear of a kid to help the kid start to nurse. If the kid has trouble finding the teat or is not nursing, place baby at teat and tickle his rear end. If the doe is not allowing the kid to nurse or doesn't have a good milk supply, you should be prepared to bottle feed the kid.

While mom is cleaning or nursing, give her a bucket of molasses water to help re-energize her. Birthing is a difficult task and takes a lot of energy from the doe. Molasses water will help her recuperate quicker. You may also want to place some feed close by so she can eat without having to go far. If you had to assist in delivery, you should start a penicillin treatment for the doe. You can also give the doe a Selenium and B1 dose as well.

The doe and kid should stay confined together for the first few days to ensure a happy bonding. You will notice that the kid's stools start as black tar and then turn to a yellow mustard color the week after birth. By the third day you can let mom out of her pen area for a few hours to browse if you have available pasture. Once the kids are over a week old, we allow them to wander around with the herd.   They quickly learn who they can go around and who they should stay away from. We still feed mom separate from the herd for another week to ensure she is getting the right amount without worrying about other goat bullies. We also add probios to her feed to help readjust her rumen.

Worming and vaccinations are issues all goat owners face. Some goat owners will tell you to worm newborn kids and the doe; others tell you only worm when needed. We do the latter. This is a decision you will need to make for your herd. In our program we only worm if testing shows a need to do so. But if you are going to treat as a precaution, kids can be safely wormed at two weeks old. If you use a vaccination program, kids should start their CD&T vaccine at four weeks, then again at weeks eight and twelve.

Although pregnancy and kidding is stressful, it is a very special event that you can be a part of. There is no other experience that compares to being there for the birth and raising kids from newborn to adulthood. We will discuss diseases and illnesses of pregnant does and kids in the Disease and Illness chapter.

Chapter Six – Bottle Babies

<u>What to bottle feed</u> – It is important for a newborn kid to receive colostrum (Mother's first milk) for the first 24 hours of life, if at all possible. After that use fresh goat's milk from another doe, possibly one who lost her kids or purchased if available. If you cannot get fresh goat's milk, use whole cow's milk from the store, not fresh cow's milk as it could transmit other diseases to your goat such as Johne's or CAE. You can also use whole canned milk diluted half with water, **not** filled evaporated milk which contains soy.

It's a good idea to add a pinch of baking soda and a pinch of salt to a bottle. Baking soda helps maintain the natural flora of the rumen and salt encourages drinking. Whatever you decide to give, stick with it. To prevent gut distress, do not change from cow's milk to canned milk or vice versa. Prior to feeding you will need to warm the milk – not hot – not cold – a little warmer than tepid – lukewarm. Test the temp on your wrist, or take a drink. It should not feel cold or hot, but warm.

It should be mentioned that there has been a lot of fear around using milk replacers on kids. I for one had always heard never to use it as it would kill the kid. Thanks to some friends, I learned that milk replacers, although not ideal, will not kill your kids. And if used properly, healthy goats can be raised using the milk replacer. The problem occurs when it is not used properly – too watered down, not enough water, not warm enough, too hot, etc. The mixing directions for milk replacers don't adjust the ratio of water/powder when it may be needed to do so. If you are thinking about using milk replacers, please do the research needed first, so that you are comfortable that they are mixed properly.

Kids and Hay – Always keep fresh hay around from birth for kids to nibble on even while bottle feeding. They won't really eat the hay, but the enzymes in the hay will help develop their rumen.

How often to bottle feed – Try to feed similar to what the doe would do. The first few days does allow the kids to eat several times throughout the day but in small amounts each time. As the kids grow, does will nurse less often, thus teaching the kids to eat more per feeding. Fill your bottle with a measured amount of milk, allow kid to nurse until the stomach feels full but not tight. Measure the remaining milk, subtract the amount that was drunk; your next feeding amount will start there. As the baby grows it will need more milk less often. Do not overfeed a kid per feeding – allow them to feed until full only. If the kid is trying to gorge himself, you are not feeding often enough. A good estimation is 15% of kids weight fed slowly over a 24hr period.

The following table is a guide for bottle feeding.

	Mini-Breed	Full-Size Breed
Day One	1-3 oz every 2-3 hrs	2-4 oz
Day Two	2-3 oz 8-10xd	3-4 oz
Day Three	3-4 oz 8xd	4 oz
Day Four	4-5 oz 7-8xd	6 oz
Week One	4-5 oz 5-6xd	6-8 oz
Weeks Two-Three	4-6 oz 4xd	6-8 oz
Weeks Four-Eight	6-8 oz 3-4xd	10-12
Two-Three Months	6-8 oz 3xd	10-12

Choosing the right Nipple – This is usually a matter of your preference and not the goats as they will adapt to whichever is chosen. We recommend using regular baby nipples and baby bottles. They are less expensive and usually have measuring amounts imprinted on the side of the bottle. Pritchard teats are made to screw onto a bottle such as a soda bottle. Lamb nipples are made to squeeze over the top of a bottle such as a beer bottle. Lamb nipples and baby nipples' holes most likely will need to be enlarged to ensure proper flow. But opening it wide could choke the baby. Pritchard nipples have no hole and the very top will need to be cut off. This is just a trial and error until you find the right size.

Getting your kid to accept the bottle – Make sure the milk is warm. Hold the bottle with thumb and forefinger circling around near the bottle end and remaining fingers wrapped around bottom of nipple area. Tilt the kids head upwards. A kid's four-chambered stomach is not yet functional. Their head has to be tilted upward to close off the rumen and allow the milk to flow correctly. Hold the bottle securely at a 45º angle downward towards the kid's mouth. Gently open the kid's mouth and insert the nipple, squeezing just a drop of milk onto the tongue as you do. Have patience as this is not natural for the kid; he's knows it's not natural and most likely will not want to feed in this manner. It takes time, but don't get discouraged and DON'T give up. Your kid has to eat now. Help him realize that it is food and it is good for him. If you have trouble getting him to eat, tickle his rear. You may have to try several times, repeating over and over. Allowing him to skip a meal encourages overeating at the next meal and will cause gut distress. Be patient, and keep at it.

Watch the kid's poop after feeding. Diarrhea means too much milk and no poop can mean gut distress. Keep enemas and CD Anti-toxin **(NOT the same as CDT)** on hand in case needed.

Weaning and Changing to Feed – Kids can be weaned naturally by their mom if dam raised. You can also wean kids on your own by six weeks of age. Some goat owners do not wean their kids until two months or older. There is no right or wrong way or timeframe for weaning. It is a choice every goat owner needs to decide. Natural weaning will occur when the doe wants the kid weaned or when she is bred again.

We start to wean our kids at four weeks of age. This is when we start to introduce feed and water. Do not just stop bottle feeding and go straight to feed/water, as this will cause gut distress and other health issues. From four to eight weeks we slowly introduce feed until kids are fully weaned. Starting at four weeks, we move our kids into the kidding pen where fresh water, hay and browse are available 24/7. We provide ¼ cup of feed per kid for the first two weeks; 1/3 cup of feed per kid at 7 weeks; up to ½ cup of feed at eight weeks.

Other breeders may wean differently or other breeds may have different requirements. Some breeders do not wean until 4 months of age. Always remember, if you notice diarrhea cut back on any new feed. It's best to take things slow when introducing anything new into a goat's diet. Sudden changes can cause many health issues.

Chapter Seven – Meat Goats

Although any goat can be considered as a 'meat' goat because they all can be consumed, there are a few breeds that are specifically used as meat goats. You should choose a goat breed based on your herd requirements and also on the individual goat selection because not all goats produce the same within a breed category.

If you have chosen to raise meat goats, you can sell the goat itself as a meat or as a pet goat, sell the goat meat if allowed to do so in your area, and you can show your goats. There is a strong market for meat goats. Many cultures use goat meat as their traditional food. You should become familiar with various cultures and their traditions and plan your selling strategies based on the need of the expected client. In this chapter we will discuss recognized meat breeds, options for selling, and some cultural requirements for goat meat.

Meat Breeds - Although any goat can be used for meat, there are a few breeds that are specifically bred for meat production.

Angora goats are a dual purpose breed for meat production and for Angora wool. They originated in turkey and although they can be raised in cold or hot climates, they are not a very hardy goat. Because of their thick fur, Angoras need to be shaved every six months.

Boer goats are the most popular meat breed. They were first developed in South Africa and are easily recognizable by their characteristic white body with a red head, convex face, Roman nose, and massive stance. Boer goats are the largest breed and are in high demand as a meat goat due to the fast growth rate. They can gain half a pound a day.

Kiko goats were first developed in New Zealand as a crossbreed of Nubian, Saanan or Toggenberg. They are usually all white and are very hardy. Kikos seem to have a natural resistance to parasites and hoof rot. Although they can wean out more pounds per kid than Boers, Boers are more popular at barn sales.

Mytonic goats are a meat breed with a special characteristic. When startled, they seem to faint or stiffen up. They are a native goat to the United States. They are quickly becoming a favorite of breeders because of their unique fainting ability. They are well-muscled in the rump and chest areas and come in a variety of colors.

Pygmy goats are a dual purpose breed raised as a meat goat but also considered a dairy breed, although they are the smallest of the goat breeds. They were originally called the Cameroon Dwarf from Africa and came to America in the 1950's. Pygmies have a meat style body, compact and well muscled. They come in a few colors but have breed-specific markings. True Pygmies are easily recognized by their body shape which is almost as wide as they are long. Does look to be permanently pregnant even when they are not.

Spanish goats were brought to America in the 1540's by Spanish explorers. They are very hardy, make great foragers and are great mothers.   However, because the Spanish goat was a wild goat, it does not have a consistent growth rate within the breed. Selection must be done on the individual goat and not on the breed itself. Spanish goats come in a variety of colors.

Savannah goats are a relatively new breed to America as they were imported in the 1990's. Their body style resembles that of the Boer breed but is usually all white with a few black hairs on their ears.

Marketing

Unless you are only raising goats for your personal use, marketing is the end reason for breeding. You need to know what your options are and where to go to sell your goats. There are basically six avenues of marketing meat goats.

(1)Live market or Auction sales allow goats to be sold any time without the need to breed for specific holidays. Owners who sell at open market get a set market price for the goat, whereas auction sales prices vary according to the bidder. With live market or auction sales, the owner can miss opportunities for higher profit.

(2)Farm sales allow the owner to control the price and time for sale of the goat. Owners can reach a higher profit margin by selling from their home, but doing so can increase the likelihood of bringing in diseases to their herd by allowing others to come on their land.

(3)Brush Control sales are an opportunity to rent out or lease goats to others in need of land clearing. The owner can rent out as many goats as needed, but then can also increase profit by choosing to slaughter and use another avenue for sales when requests for brush control diminishes. This method also increases the chances for disease and illness when goats are brought back to the owner's property.

(4)Breeding Stock sales allow the owner to breed for commercial herds or show herds. Higher quality goats bring in more profit, whereas lower quality stock can be sold through another avenue. This sales method is time consuming in the fact that the owner must continually breed up his stock to improve line quality.

(5)Pet and home sales are another avenue an owner can use for goats that do not fit in with one of the other sales methods. This avenue is an increasing demand for Pygmy goats as more consumers look for the backyard goat for personal home use.

(6)Direct Meat sales usually require certification of your facilities by a government agency. Once approved, the owner can slaughter the animals and sell the goat meat directly to stores or consumers. This avenue usually incurs higher upfront costs and restrictions based on local and state regulations.

Goat meat is a healthy alternative for meals. It has just about the same amount of calories as chicken, but is lower in fat and higher in protein and Iron.

3 oz. Cooked	Calories	Fat (g)	Sat Fat (g)	Protein (g)	Iron (mg)
Goat	122	2.58	0.79	23	3.3
Beef	245	16	6.8	23	2.9
Pork	310	24	8.7	21	2.7
Lamb	235	16	7.3	22	1.4
Chicken	120	3.5	1.1	21	1.5

Marketing for Holidays – Goat meat sales increase around specific cultural holidays. It is important to know the holidays and goat requirements of various ethnic populations to meet the demand. Many cultures have special slaughter requirements that must be observed and requirements for how the goat was raised. Also, some cultures prefer dark colored or black goats while many avoid all white goats, Pygmy goats or Angora goats. Boer goats seem to be popular with most cultures.

Caribbean holidays occur in the fall. Intact market kids or bucks are wanted during this time . But castrated kids and females are becoming more popular as well. For meat sales, the curried goat and goat soup are popular dishes.

At Christmas and New Year 's suckling kids are most popular. To meet this demand you would breed does for fall kidding. However, this is a breed season outside the normal breeding season for most breeds. Goat stew is a popular dish for Latin cultures during the Christmas season.

Cinco de Mayo and other Hispanic holidays need kids and yearlings for celebrations. Whole goat barbeque and roasts are traditional dishes during these times.

Dassai is a Hindu holiday where does are not acceptable for feasts. There is a preference for market wether kids with varying weight sizes depending on the number of people to feed.

Festival of the Sacrifice is a traditional Muslim holiday requiring yearlings or large market kids and older goats. The goat must be healthy with no cuts, scars, or wounds and should not be castrated.

Ramadan is a month long holiday for Muslims. The preference during this time is for market kids between 45 to 110 pounds. The goat cannot have any adult teeth.

Roman or Western Easter market is a traditional Italian market. Kids between 18 and 35 pounds and heavier are needed during this time. This would include suckling kids, market kids, and older goats. For Greek or Orthodox Easter, it is usually 25 to 50 pound suckling kids.

When marketing your goats for sale, the owner should become familiar with marketing terms. Raising goats in a specific manner or a specific slaughter method can increase sales and open up other avenues for sale previously not used.

'Grass Fed' is a goat that has been raised at the minimum 80% of natural pasture or forage with little grain or feed additives.

'Humanely raised' means a humane animal care program was followed, and requires certification through an agency. Most require the animal has natural shade and humane slaughter.

'Natural' is a term referring to how the goat was slaughtered or processed. The goat cannot have any artificial ingredients, colorings, or chemical preservatives.

'Organically Certified' requires goats to be fed organically grown feed and must comply with specific natural guidelines. Most often, antibiotics and wormers are not allowed to be used on any goat considered to be organically certified.

Many religious cultures also have special requirements on how the goat is to be slaughtered. Most Muslims require a *Halal* slaughter by an adult Muslim using the *zabiha* method. Jewish cultures require a *kosher* slaughter by an Orthodox Jew.

If you are going to sell meat goats you should also be familiar with the market grades. Currently there are three known grades of 1, 2, or 3. Grade 1 is a very well-muscled goat with thickness around its hind legs, a full well-rounded backstrap, and thick shoulder areas. Grade 2 is an average moderately muscled goat, whereas a Grade 3 is an underweight goat with very little muscle.

Raising meat goats can become profitable if you research your options and become familiar with cultural and legal requirements. As with everything else, know all the details before you decide to take on the challenges of marketing.

Chapter Eight – Dairy Goats

The use of goats for dairy, meat, hair and skin has been around for thousands of years. Today, goats are once again becoming a favorite for families - especially for those who have allergies to cow's milk products. Goat's milk is easier to digest and has more vitamins than any other milk.

	Goat	Cow	Human
Protein (percent)	3	3	1.1
Fat (percent)	3.8	3.6	4
Calories	70	69	68
Vitamin A	39	21	32
Vitamin B1	68	45	17
Riboflavin	210	159	26
Vitamin C	2	2	3
Vitamin D	0.7	0.7	0.3
Calcium	0.19	0.18	0.04
Iron	0.07	0.06	0.2
Phosphorus	0.27	0	0.06
Cholesterol	12	15	20

Dairy Breeds - There are eight recognized American Dairy Goat Association (ADGA) dairy breeds.

Alpines originated in the French Alps, and come in most colors with the exception of all white and light brown with markings. They are considered a medium to large breed and they are the leading breed for dairy production.

Lamanchas began in the USA in Oregon from crossing short eared goats with a Nubian. They have very tiny to no outside ears and are considered to be a small breed. Lamanchas are relatively calm and quiet. They come in a variety of colors.

Nigerians are a mini-sized dairy breed. All colors are acceptable for this breed. They have the same body conformation as other dairy breeds, but produce less milk than their larger counterparts. Their milk is well known for its richness and high butterfat content.

Nubians are the most popular breed of dairy goats. They originated from a crossbreed of English dairy goats with a lop-eared goat. They come in many colors and have a very distinctive high pitched call. Nubians can be stubborn at times, more so than other goat breeds, but produce milk which is high in protein and butterfat.

Oberhaslis or Swiss Alpines have very specific coloring and markings. They should be a chamois color with a black dorsal stripe, udder, belly, lower legs and a nearly black head.

Saanens are another Swiss breed and should be solid white, although a light crème color is acceptable. They are the tallest goat breed. Sables are the colored version of Saanens.

Toggenburgs are the oldest registered breed. They have specific coloring from a light fawn color to a dark chocolate brown with white ears and lower legs, white stripes on tail and face. Toggenburgs are high spirited and relatively small.

Dairy goats have the same marketing avenues as do meat goats and also the possibility of selling their milk and dairy products. Learning how to milk a goat is similar to learning how to milk a cow. It can be done by hand or with a milking machine. Either way it should be done in a sanitary manner. By hand, milking will take some time to learn, but keep at it. Practice makes perfect. A dairy goat not feeding a kid should be milked out at least twice a day.

How to milk a goat - You should have a milking stand at least a foot off the ground for the goat to go up on while milking. This allows the goat to be at proper height level as you sit to milk. Prepare a food bucket, no more than a pound of feed, for the goat to be milked. Get the goat up on the stand and place the feed bucket in front of the doe.

If you haven't already, shave the doe's udder area and then wash the udder and teats with warm water. The warmth relaxes the doe and sends the signal for milk to let down. Place your milk bucket below the doe under the teats. Sit down on a bucket or chair to the back side of the doe.

Wrap your thumb and forefinger around the top of the teat to close off the milk in the teat from the udder. For the first let down, point the teat away from your milking bucket. With slight pressure close your middle finger, then ring finger, then pinky finger around the teat in succession - 1,2,3, keeping your thumb and forefinger closed. Do not pull down on the teat. With the teat aimed away from your bucket, the first milk should spray out. Relax your thumb and forefinger to allow more milk to let down. Continue through the steps for milking rotating to each teat, but aiming into the milk bucket. Stop milking when there is less milk coming out or when the teats look deflated.

Gently massage the udder to clear the teats. Remove the milk bucket. Use a teat dip or a warm wash with Iodine or Betadine to wash the udder and teats. Allow the doe to return to the herd.

I always recommend pasteurizing the goat's milk, but if your doe and herd is disease free you can drink or use the milk unpasteurized. Pasteurizing your milk kills off bacteria that normally could cause illness. To pasteurize your milk you can purchase a milk pasteurizer or do it yourself.

What you will need to home pasteurize milk: A candy thermometer, cheese cloth, stainless steel pot, metal spoon, bowl (larger than pot) of cold water and ice, and glass jars.

Clean and sanitize all items. First, strain milk through cheesecloth into the pot. Place thermometer onto pot or dip it often into the milk to check temperature. Set the pot on stove on medium high, stirring the milk continually. Watch the temperature of milk. When it reaches 165°, cook for another 15 seconds. Place the pot into the bowl of icy water and stir milk. When the ice has melted, pour the milk into the glass jars and immediately place them into the freezer. Once the milk has cooled in the freezer, you can place the jars into the refrigerator. Shake the milk to mix prior to drinking.

You can also use goat's milk to make a variety of meals and desserts; just use it in place of cow's milk. Soaps and lotions and other body products can also be made with goat's milk. These products can then be sold for additional income. There is so much that a goat can be used for – maybe the reason why they are becoming so popular again.

Chapter Nine – Disease and Illness

Although goats are pretty hardy and remain healthy with good management practices, there are times when your goat may become ill. It helps to know a little about some of the illnesses and diseases that can affect goats. In this chapter we will discuss some of the more common diseases and illnesses and their treatment options. <u>All medicines and dosages listed are for reference only and are not to be considered as veterinarian advice.</u> Dosages listed are those we have used in our herd(s). It is important to find a qualified veterinarian that is familiar with goats **before** one is needed.

Parasites - The biggest issue facing goats are parasites - Mites, Lice, Bot Fly, Ticks, and Fleas. There are several species of mites that can infect goats with the most common being:

- follicle mite aka demodectic mange which occurs most in young animals, pregnant does, and dairy goats. Papules usually appear on the face, neck, maxilla region, or udder
- scabies mite aka sarcoptic mange – usually resolves without treatment but heavy infestations progress showing crusty lesions and extensive hair loss around the muzzle, eyes, and ears; lesions on the inner thighs extending to the hocks, underside, and axillary region; dermal thickening and wrinkling on the scrotum and ears; and dry, scaly skin on all parts of the body.
- psoroptic ear mite lesions cause crust formation, foul odor ear discharge, and change in behavior such as scratching

the ears, head shaking, loss of equilibrium, and
spasmodic contractions of neck muscles.

- chorioptic scab mite aka chorioptic mange or leg mites
occurs on the legs and feet.

Goat lice are host specific and not transferable to humans or
other animals. There are two classes of lice – sucking or
biting. Louse-infested animals usually have a dull, matted coat
and show excessive scratching and self-grooming behavior.

Although not a burrowing or sucking parasite, the Bot fly will
deposit its larvae or maggots into a goat's nose. Symptoms
include runny nose with blood flecks in the nasal discharge,
excitability, snorting with noses towards ground.

Ticks are either one-host, two-host, or three host specific
which means their lifecycle is carried out on either one, two or
three separate hosts. Unfortunately the most common goat ticks
are usually the three host ticks which expand the infestation
within a herd and are more difficult to treat.

Goats are most often infected by cat fleas or Sticktight
flea. Cat fleas are usually found throughout the hair, but the
Sticktight flea usually stays near the head, face and ears.

Treatments

Fortunately, treatment and prevention is the same for all
these parasites. Clean out all bedding areas and replace with
each treatment. *Most livestock medicines are not shown as for*

use on goats and are used off-label. The dosages we provide on treatments are from our personal experience only. Please always consult a veterinarian to ascertain the proper medication to use and its recommended dosage.

Option 1) Invomec or ivermectin pour on – Always wear rubber gloves as this is poison – For goats under 25 pounds, dip a cotton ball in a small bowl of the liquid and run the cotton ball down the spine starting at the back of the top of the head. Goats over 25 pounds, do the same as above and repeat three times. Repeat in 10 days for three doses to ensure all live parasites, eggs and nits are killed. (This is the option we use).

Option 2) In the warmer months this is a great solution – Borax and Hydrogen Peroxide have long been known to have benefits for mange/mite treatment in dogs and other animals. It has also been used in the goat world with great success. Prepare a 5gal bucket of warm water with one box of Borax laundry detergent (save 2 cups) and one bottle of hydrogen peroxide (save 1/2 cup). Wash the goat completely, saturating the entire body, especially around hoof area. Prepare a Borax/Peroxide paste by combining the saved 2 cups Borax and 1/2 cup Peroxide. After the above wash do not rinse; now apply the paste to the most infected areas. Repeat this treatment once a week for two months.

Option 3) Cat flea and tick powder for adult goats – sprinkle from head to tail and rub down to skin; Use Kitten flea and tick powder for kids. Repeat every 10 days for three treatments.

Option 4) Cat flea and tick shampoo for adults – Bathe as normal, repeat every 7-10 days; Use Kitten shampoo for kids.

Option 5) Non-chemical – Natural products:

A) Farnam Equisect – natural oils of citronella, clove stem, thyme and corn mint. Just spray over goat and rub in.

B) Oil and Herbal Repellent Recipe

4 cups apple cider vinegar
4 cloves garlic
4 teabags of black tea
3 cloves

Place in pot and bring to a boil, turn down and simmer on low for 10 mins, set aside to cool, then steep overnight in refrigerator. Take out and strain solids from the liquid. Add the following:
2 tsp chrysanthemum oil
2 tbsp dish soap
1 cup Avon skin so soft
2 tsp neem oil
2tsp Tea Tree Oil

Add all to spray bottle. Spray all over goat. It can be used weekly as preventative or daily as treatment for fleas, ticks, mites, bots, or lice.

Worms - The second biggest issue with goats are worms. Wormers are becoming ineffective for treatment due to overuse. Please always have a fecal test done to determine what type of worm is present to know which wormer product to use. Again, most livestock medicines are not shown as for use on goats and are used off-label. The dosages we provide on treatments are from our personal experience only. *Please always consult a veterinarian to ascertain the proper medication to use and its recommended dosage.* Elwood Ranch does not support, recommend, or favor any particular medicine.

Y = yes   N = no   U = unknown   d=days

Med Name	Gen Name	Lung Worm	Round worm	larva	lice	mite	Preg Safe	Milk Hold
Ivomec 1%	ivermectin	Y	Y	Y	Y	Y	Y	14d
Double Impact 1%	ivermectin	Y	Y	Y	Y	Y	Y	14d
Dectomax	Doramectin 1% or .5%	Y	Y	Y	Y	Y	U	36d
Eprinex	Eprinomectin 5mg	Y	Y	Y	Y	Y	U	N
Valbazen	albendazole 11.36%	Y	Y				N	5d
Safeguard	*fenbendazole*	Y	Y				Y	4d
Panacur	*fenbendazole*	Y	Y				Y	4d
Tramisole	levamisole	Y	Y				Y	4d
Levasol	levamisole	Y	Y				Y	4d
Cydectin	moxidectin	Y	Y	Y	Y	Y	N	N
Synanthic	oxfendazole	Y					N	N

Med Name	Gen Name	stoma worm	Tape worm	Intes Worm	liver fluke	Preg Safe	Milk Hold
Ivomec 1%	ivermectin					Y	14d
Double Impact 1%	ivermectin					Y	14d
Dectomax	Doramectin 1% or .5%					U	36d
Eprinex	Eprinomectin 5mg					U	N
Valbazen	albendazole 11.36%	Y	Y	Y	Y	N	5d
Safeguard	fenbendazole	Y	Y	Y	Y	Y	4d
Panacur	fenbendazole	Y	Y	Y	Y	Y	4d
Tramisole	levamisole					Y	4d
Levasol	levamisole					Y	4d
Cydectin	moxidectin					N	N
Synanthic	oxfendazole	Y	Y	Y		N	N

Brand Names: Ivomec paste 1.87% **Generic Name:** Ivermectin

Goat dose: A full tube is usually 6cc. Triple your goat's weight and then dose as follows - 1cc/200lbs; 1/2cc/100lbs, 1/4cc/50 pounds given orally.   Repeat every 10 days for three doses.   Milk withholding time: 14 days per dose. Ivomec/Invermectin are safe for use in pregnant does.

Brand Names: Ivomec injectable for Cattle & Swine 1% Sterile Solution (Merck); Double Impact Injectable for Cattle & Swine 1% Sterile Solution (Agrilabs) **Generic Name:** Ivermectin

Goat dose: **Do not inject this wormer.** It is given orally on a goat. 1ml per 50 pounds- given orally. Milk withholding time: 14 days. Ivomec/Invermectin are safe for use in pregnant does.

Brand Names: Dectomax – injectable & pour-on; **Generic Name:** doramectin 1% injectable, doramectin 0.5% pour-on

Goat dose: **Do not inject this wormer.** It is given orally on a goat. Pour-On: 1 ml per 10 lbs – given orally, Injectable: 1 ml per 35 lbs – given orally; Milk withholding time: 36 days

Brand Names: Eprinex; **Generic Name:** eprinomectin 5mg

Goat dose: Pour-On: 1 ml per 10 lbs; Milk withholding time: None

Brand Names: Valbazen Cattle & Sheep Dewormer Suspension (Pfizer) **Generic Name:** albendazole 11.36% (This is a brand we use)

Goat dose: 1 ml per 10 pounds- given orally; Milk withholding time: 5 days; not safe for pregnant does.

Brand Names: Safeguard, Panacur **Generic Name:** fenbendazole (Not very Effective due to overuse in USA)

Goat dose: If using the horse or cattle version, give 4 times the recommended dose orally; treat for tapeworms – three days in a row; Milk withholding time: 4 days; Safe for use in pregnant does.

Brand Names: Tramisole, Levasol; **Generic Name:** levamisole (Death occurs when overdosed)

Goat dose: Injection of Tramisol 13.65% is given SQ at the rate of 2ml/100lbs; Oral: 2 tablets/100 lbs. Milk withholding time: 4 days; Safe in pregnant animals.

Brand Names: Cydectin; **Generic Name:** moxidectin (DO NOT USE AS POUR ON)

Goat dose: Cydectin- given orally 1cc per 20-25 lbs; Not safe for pregnant does; keep away from skin.

Brand Names: Synanthic, Benzelmin (horse) **Generic Name:** oxfendazole

Goat dose: Oral 2x – 3x the label dosage To treat for tape worms, you need to give the above treatment three days in a row, Do not use in pregnant animals.

Meningeal Worm, Paralaphostrongylus tenius, is an internal parasite usually affecting the white-tailed deer. At times it can infest a goat and cause severe neurological problems including fatality. Usually a goat is infected by eating a snail or slug who acts as an intermediate host. Once a goat is infected by the parasite, it quickly moves into the brain and spinal cord. Unlike in deer, the meningeal larvae infesting goats do not mature into adults. They go throughout the goats central nervous system, damaging nervous tissue causing the neurological symptoms.

Symptoms

It usually takes 10-14 days for symptoms to begin showing after a goat has ingested an infected snail or slug. This is the time it takes for the parasite to reach the brain and spinal cord. Symptoms vary depending upon the number of parasites within the goat. Mild symptoms start with a slight limp or weakness in legs, head tilting, circling, lethargy, lack of interest in food. However, in most cases, infected goats remain alert and continue to eat and drink normally.   As it progresses or more parasites reach the spinal cord and brain, the goat may become partially or completely paralyzed, and may experience some blindness. Other diseases can mimic the symptoms of Meningeal infection such as CAE, Listerosis, Rabies, Scrapie, Polioencephalomalicia, copper or selenium deficiency, and spinal or brain abscesses.

Diagnosis

Unfortunately, the Meningeal worm cannot be diagnosed through fecal examination because the larvae do not produce eggs and are not passed through fecal material. They also cannot be diagnosed through blood tests. The only way for an accurate diagnosis is after death by necropsy testing. It has been shown possible at times to diagnose by testing the cerebrospinal fluid in a live animal, but this may result in death of the goat and is not completely accurate. Most diagnosis is based on symptoms and history of the goat when the goat has been allowed to graze on pasture and there is deer in the area.

Prevention

Prevention is the best management practice for possible meningeal parasites. Goats in areas known to have a high deer population should not be raised on pasture, or removed from pasture during rainy season and cold weather. Goats should only be allowed to browse on high and dry land and in areas with little to no woodland and restrained from occupying areas with ponds, swamps, wetlands, or low-lying, poorly-drained fields. To test known deer areas for the presence of snails and slugs, place dry dog food on the top of a plastic potato chip lid on the ground and cover it with a box. Check the food at dawn and at dusk. If snails or slugs are seen, there is a possibility of Meningeal worm infestation.

Treatment

High doses of Ivomec Plus should be given at a rate of 1cc/25# for ten days; Dexamethasone at 1-2mg/20# 1xd for three days – Use prednisolone sodium succinate for pregnant does instead of Dexamethasone at 0.5mg/20# SC, 1xd for three days; after Ivomec treatment dose 4.6 ml Safe-Guard per 100# for five days. Also add in a B1/B12, Vit E, and Vit A regimen during treatment and for one month after. All goats within a herd should be treated at the time of possible infection.

Treatment will not improve damage already done, but will prevent further symptoms from developing. The earlier treatment is started, the better the outcome. During the high risk season from November to February, it is recommended to use Ivomec as a monthly preventative.

General Colds - Symptoms: runny nose with clear to white snot, cough, no temperature.

Treatment

Maintain a clean housing environment, well-ventilated and draft-free. Give probiotics. Do not give antibiotics when there are only cold symptoms. Antibiotics wreak havoc on the digestive system in goats.

Symptoms of Other than Cold – Goats can be very prone to Pneumonia. Symptoms can be runny nose with green snot (not cud or grass), raspy lungs, temperature.

Treatment

Maintain a clean housing environment, well-ventilated and draft-free. Give probiotics. Give aromatherapy or a vaporizer with Eucalyptus, Tea Tree, Lavender, and Thyme essential oils, or Vick's Vapor Rub. Oxytetracycline (if the goat is not pregnant) **OR** Penicillin (if goat is pregnant).

Cough – Usually a sign of lungworms.

Treatment: Use Ivomec de-wormer.

Sneezing – Usually nothing to worry about. Allergies, dusty or moldy hay, or alfalfa hay may cause sneezing. Goats will also sneeze during play or to sound an alarm.

Lumps in Goats – Milk Neck, Bottle Jaw, Caseous Lymphadenitis, Vaccines

Milk Neck occurs on a kid goat and is a soft swelling/lump on the chin or throat where the chin and throat meet. Milk Neck occurs from a kid milking on a heavy milk producing doe. There is no treatment and the lump/swelling will go away on its own after the kid is weaned.

Bottle Jaw usually occurs in adult goats and not kids. It is a soft swelling/lump on the jaw or chin. The Treatment: Worm the goat immediately. If not treated, the goat could die.

Caseous is a hard lump, usually about the size of a quarter located near a lymph gland. Any hard lump should be inspected by a vet.

Vaccination lump is a hard lump near an injection site. There is no treatment. The lump should go away on its own, but could take up to a year to disappear.

Goat Bloat

A goat's rumen (stomach area) is a big fermentation vat. The bigger it is the healthier they process their food. Look at their belly. Look straight down your goat from the front head. Is their width side to side big and evenly big on both sides?   That "fat" is good rumen development and a sign of a healthy goat. Now with Bloat, one side (usually the left) of the goat will be wider than the right. Bloat can cause difficulty breathing.

What causes Goat Bloat?

Overeating, a sudden change in a goat's diet; eating certain weeds such as Milkweed; giving a goat grass hay or hay that is still wet or moldy; obstruction of the esophagus; Tetanus or face paralysis; and in kids it may occur when given milk replacers instead of goat's milk.

Signs of Goat Bloat:

The goat's stomach protrudes out more than normal, mostly the left side (rumen area) will look bigger; your goat will show signs of discomfort such as kicking, 'mawing', or grinding their teeth (not to be confused with normal "cud chew"); your goat may have a lack of interest in normal activities or seem depressed; and in severe cases difficulty breathing.

Prevention of Goat Bloat:

1. Provide goats with fresh, good quality hay and restrict grazing time on rapid growing pastures.

2. Provide baking soda (sodium bicarbonate) free
    choice. Baking soda aids in balancing the pH level in the
    rumen and helps to keep the digestive processes in tune.

3.   Always make changes in diet gradually.

Goat Bloat Treatment:

•   Stop the goat from eating.

•   Give your goats 1/4 – 1/3 cup of vegetable/peanut oil
    orally (NOT mineral oil). The oil breaks the bubbles in the
    stomach so they can poop and expel the excess gas.

•   Massage goat's sides, especially the left side (rumen)
    until the goat begins to burp and fart.

If the bloat is really bad or your goat has trouble breathing,
call a veterinarian immediately! The pressure in the stomach
can stop the lungs or heart from working. The veterinarian can
release the gas by lying the goat down on its right side and
making a small incision behind the bottom of the ribs on the left
side of the goat.

Goat Stones - Urinary Calculi, Goat Stones, Urolithiasis

The first lesson I learned and will never forget was with this
very common, preventable, and treatable disorder effecting
bucks and wethers. I purchased a polled wether from a zoo and
knew nothing about goat stones and how easily wethers could

get them. Joshua was the friendliest goat I have ever had and was my favorite pet. I spoiled him rotten and he would follow me everywhere. We had him for several months before he got sick.

At first he acted like it hurt to pee. I researched online and learned about Urinary Calculi (UC). From my research I learned I was not feeding him a proper diet for a wether/buck which is the main cause for developing goat stones. So I added more hay to his diet, added Ammonium chloride and salt to his feed. But then he stopped wanting to eat and drink. I checked him and noticed the tip of his penis was encrusted. First mistake, I pulled whatever it was off. I called many veterinarians and finally found one, the only one I knew about at that time, which treated goats. My second mistake, I took him to this vet who ended up not knowing enough about goats.

The veterinarian poked and prodded poor Joshua who was a champ and allowed him to continue even though you could tell he was in pain. The vet told us yes, he believed it was UC and would have to drain the urine. My thought from what I read online was he would remove the vermiform appendage and then catheterize him to remove the excess urine. Before I knew what was happening, the veterinarian took a syringe and needle and stuck it into Joshua's belly. He couldn't get anything out and said he must have missed and did it again. Poor Joshua was screaming his head off. I yelled at the vet, got him to remove the needle and hauled my poor Joshua off the table and home.

I can't say for certain, but it is my belief that the vet ruptured his stomach or bladder or wherever it was that he stuck that needle through. By the time I got him home, Joshua just laid down and was crying a lot. He would not move. I lay down with him and he would try to lift his head up and look at me and the look I got was such agony and sorrow. He was in so much pain and there was nothing I could do to help him. My husband finally put him down to end his suffering. RIP my love, My Joshua, you will forever remain in my heart.

Now that my sob story is over, here is what I have learned about UC, how to prevent, and how to treat.

Note: Common name terms: distal flexure = sigmoid flexure; vermiform appendage = pizzle = urethral process

Bucks and especially wethers are susceptible to UC. Proper diet is essential in preventing UC, and hay is an essential part of that diet. Grain feeds need to be monitored for the proper calcium to phosphorus ratio which should be 2 to 1. Stones begin to form when the urine ph levels cause minerals to bind to form crystals in the urinary track, similar to kidney stones in humans.

The stones block the flow of urine causing great pain, discomfort, and oftentimes death. This is because stones can become lodged in the bend in the penis, called the distal flexure, or at the small appendage at the tip of the penis, called the vermiform appendage. Wethers are more prone to UC usually because they are often castrated at too young of an age. Castration stops the growth of the urethra. If done before the urethra has been given ample time to grow to its normal diameter, it becomes more difficult for the solid particles to be eliminated.

Symptoms

Many symptoms of goat stones mimic the symptoms humans feel when they have stones. A goat will show signs of abdominal discomfort, lethargy, restlessness; they may kick at their belly, try to pee often, dribble, or seem to be in pain when trying to pee; they may lie down often and cry; they may have no appetite; there may be drops of blood in the urine, or a crystallized mess stuck to the hairs around the penis opening, the penis itself may be swollen and/or stones may be felt in the vermiform appendage. If the urethra has ruptured, the abdomen may become swollen.

Prevention

High hay diet with proper grain feed levels of calcium and phosphorus. Adding 4% salt as a top feed to increase thirst and drinking; add 1cup Apple Cider Vinegar (ACV) to 30 gal water supply 2x week will all help prevent UC. A cranberry mineral block and/or cranberry mash may be given weekly as well. I prefer these natural methods for prevention and have not had another case of UC since.

Chemical prevention – add Ammonium sulfate as a top feed at the rate of 15g/100# goat per day; or Ammonium Chloride (AC) at 1tsp/150# goat per day.

Treatment

If goat is still able to urinate, withhold feed for 24hours. Prepare AC mix (1.5tsp AC for every 75# of goat to 20cc water) Give oral dose Day 1-3: 2xd at 12 hour interval, Day 3-7: 1/2tsp AC Mix 2xd every 12hrs. AC burns the throat so you may have to use a stomach tube to drench. Start a strict Hay and ACV/water diet during treatment for the next two weeks. **DO NOT force a goat with UC to drink water.** If there is a blockage, the water has nowhere to go.

After two weeks, slowly add the proper grain feed ratio back into the diet with AC at 1 tsp as a top feed. Start feeding with only 1/8 cup 1xd first three days, then 1/8 cup 2xd for next three days, then go to ¼ cup 2xd. Continue the ACV/water and prevention treatments listed above.

If the goat is unable to urinate or seems to be blocked, a veterinarian may be required. Treatment is usually very costly and not always effective. Stones can be lodged at the distal flexure of the penis or at the vermiform appendage. If it is lodged at or near the vermiform appendage, an experienced owner may be able to remove the appendage to provide temporary relief and catheterize the goat at home to relieve buildup of urine/stones and restore normal flow. 80% of cases with appendage removal reoccur. **REPEAT – ONLY AN EXPERIENCED GOAT OWNER SHOULD ATTEMPT THIS PROCEDURE.** Always, if possible, consult a veterinarian who knows about and is experienced with goats – immediately!

To remove the vermiform appendage, have the goat in a sitting position and manually work the penis out of the shaft. The vermiform appendage, a curly appendage at the end of the penis, must be cut off prior to catheterizing. Often the vermiform appendage is black and crusty in goats with UC. The vermiform appendage can be cut off with sterilized, small, sharp, surgical scissors. A local numbing agent should be used. Once the appendage is removed, the goat may be catheterized to release the buildup of urine and allow urine to flow normally again. I will not go into detail as to how this is done because I have no experience doing the procedure. Review our reference chapter for further information.

Goat Chlamydia and "Pink Eye"

This is a bacterial infection which is usually acute and spreads very quickly among goats. It can be spread by flies, especially during summer months. Young goats are most susceptible, but it is also a major cause of abortion in goats. A doe may show no outward signs but will abort when pregnant if not treated.

Symptoms of Goat Chlamydia or "Pink Eye": Continuous or spasmodic winking, Tearing of the eye, Discharge from the eye, Opaque eye, Inflammation of the cornea, Intolerance to light, Infectious arthritis, Mammary gland or uterine infection. It can result in permanent blindness and spontaneous abortions when left untreated.

Treatment of Chlamydia and "Pink Eye": Give all goats in the herd 3ml of 200mg Oxytetracycline injected SubQ once daily for 3 days; and two drops of Oxytetracycline directly into the eye for three days. Review Drug Facts.

Coccidiosis, Coccidia, Coccidian in Goats

Coccidiosis is very contagious and will spread through a herd like wildfire! Coccidia in goats are intestinal coccidian parasites of the genus Eimeria and is species specific. One species of animal cannot infect animals of another species. Chickens cannot infect goats; goats cannot infect dogs and so on. All goats carry a few coccidian within their intestines.

Symptoms in Goats: Coccidiosis symptoms occur when there is an overload of coccidia or worms within your goat. Does who have just kidded and young kids are the most susceptible to a coccidian overload. Adult goats usually have immunities built up but may contract coccidiosis under stress or when there is a sudden change in feed. However, because the parasite which causes the symptoms of Coccidiosis is passed through fecal-to-oral contact, all goats can become overloaded with coccidia through contamination.

Initial symptoms include diarrhea, dehydration and sometimes fever. If treatment isn't begun immediately, permanent damage will be done to the intestinal lining and the goat may die. For goats who survive, weight loss is substantial and sometimes chronic. In advanced cases of Coccidiosis, diarrhea can be watery and may contain mucous and blood.

Treatment of Coccidiosis in Goats: Prevention is the best treatment for coccidia infection. Coccidian overload is a disease caused by stress from overcrowding, dirty and/or wet pens, and unclean water. **Maintain a healthy environment, maintain healthy goats.** Kids can be put on a prevention treatment using Albon or Di-methox 12.5% liquid as follows:

Prevention – Kids at 3,6,9 weeks: Dosage given orally: Day1 – 2cc per 5lbs, Day 2/3 – 1cc per 5lbs.

For active coccidia diarrhea symptoms in kids: 4cc of Di-Methox 12.5% orally five days.

For adult treatment: 10cc orally for five days.

NOTE: During treatment it is important to remove all grain feed, increase hay availability, add probios and a B1/B12 regimen to their diet. In goats that are off feed, stomach tubing may be necessary. If the goat has severe diarrhea and/or fever, additional medications may be necessary. Please review our General Medications chapter for other treatment options. If you do not see some improvement with treatment in two days, you should consider switching to something else.

Sore mouth, Orf, Scabby Mouth, Contagious ecthyma, Pustular dermatitis, Malignant Aphtha

Sore mouth is a viral infection caused by a poxvirus (orf virus). While it is highly contagious within a goat herd it is not a major cull issue. Sore mouth is far more manageable than many other illnesses that a goat herd might encounter. <u>It is also transmittable from goat to human and proper safety precautions should be taken when treating an infected goat.</u>

Sore mouth usually runs its course within 3-4 weeks. The main concern with sore mouth is because it infects areas of the mouth, it can cause severe discomfort, and goats may stop eating and drinking. Young kids infected with sore mouth are at higher risk of death because of this and may need to be tube fed until the virus has run its course. Boer goats seem to be highly susceptible to sore mouth virus and get more severe infections.

Goats can become infected by coming into contact with the virus from shedding scabs that have contaminated bedding, feed, the ground, or by direct contact with infected animals. Vaccinating a virus-free herd will infect the herd with the virus because it is a live virus vaccine. Once a goat has had the sore mouth virus, they may build up a natural immunity to it. However, because there are different strains of the virus, it is possible for the goat to become infected again, usually several years apart from first infection. Secondary infections of the virus are usually less severe than first infection.

Humans can become infected by the virus through contact with an infected animal, usually by not taking proper safety precautions such as wearing gloves when bottle feeding, tube feeding, shearing, petting, handling infected equipment, and through an open cut or sore. Those who become infected from sore mouth usually show lesions or blisters on their hands. The infection may be painful but will clear up within two months and is not transferable human to humans.

Prevention

Quarantine all new goats for a minimum of six weeks. Do not bring a goat infected with sore mouth or a goat from a herd known to have a prior infection into your herd.

Symptoms

Symptoms begin to appear 2-3 days after initial contact with virus. At onset, blisters or pustules appear and then become thick, crusty scabs which when pulled or coming off tend to be bloody underneath. Sores are usually found on the lips, nose, and mouth, but may also appear on the ears, lower legs, genitals, and teats. Secondary infections may develop such as staph and mastitis, as well as rumen upset due to lack of proper diet during infection. In severe infections, maggot or blowfly infestation may occur. Foot and Mouth disease, Pearmouth, goat pox, and bluetongue resemble sore mouth infection, but FMD has not been seen in the US since the 30's.

Treatment

Always wear rubber gloves when treating sore mouth, preferably ones that go up to the elbow. Viruses are not treatable with antibiotics, and antibiotics should only be used to treat a secondary infection. Do NOT remove scabs from an infected goat. Doing so will increase likelihood of spreading the virus to the handler and promotes scarring. There is no cure for Sore mouth, you can only make the goat more comfortable and aid in the healing of the sores.

Apply Gentian Violet to the sores – this is found or ordered through a local drugstore. This is an ole' timer's remedy and not only helps heal the lesions but also deters secondary infections. If unavailable use an antibacterial ointment.

Vaccinating an infected herd may reduce symptoms and healing time, but has not been proven to protect a herd from becoming infected. A B1/B12 regimen should be used to help the goat stay active and help fight off secondary infections. Probios may be needed if the goat goes off feed. Tube feeding may also be required, especially in kids affected with the virus. Horse fly spray can help deter flys from infecting the goats with maggot infestation.

Show Goats – IMO if you have goats you show and have had sore mouth in your herd, it is best not to show your goats in the future. Goats can be carriers of the virus and show no symptoms but still be contagious to other goats. IMO it is irresponsible to bring a goat to a public arena where it could possibly infect other goats. With that said, shows will allow previously infected herds to show their goats if they have been vaccinated two months prior to show.

Enterotoxaemia, Overeating Disease, Pulpy Kidney Disease, **Floppy Kid Syndrome(FKS),** Clostridium Disease

I placed these illnesses here together because they are often misdiagnosed. Clostridium perfringens is commonly found within the soil and in the GI tract of goats. There are five types of this bacteria – A, B, C, D, and E (I have heard of a type F, but have found no verifier for this). All of the types cause toxins which can lead to death. Enterotoxaemia is caused by an overload of the C or D bacterium which then produces the fatal toxins.

Floppy kid syndrome is often misdiagnosed when in fact it is a case of Enterotoxaemia. FKS has been reported in some cases to be caused by the type A or E bacterium. However I have found no research to prove this and the type E is rarely found in the US. Enterotoxaemia and FKS are prevented in the same manner and treatment is the same as well. Although Enterotoxaemia has been around since 1892, the first named case of Floppy Kid Syndrome occurred in 1987. The CD vaccine was introduced as prevention against Enterotoxaemia caused by the Clostridium Perfringens bacterium type C or D.

Symptoms

Unfortunately there often can be no prior symptoms except for sudden death in the acute form of Enterotoxaemia. If you are lucky and there are forewarning symptoms of a Clostridium explosion, these could include one or more or all of the following symptoms:

Depression
No stools or diarrhea
Sloshy extended gut
lethargic
Inability to suckle or overgorging
Wobbly or unsteady gate
lying down on side a lot
Abdominal pain
unable to stand
unable to hold their heads up
fever
respiratory distress
convulsions

FKS usually occurs in kids from 3 days old to 10 days old. The reason for its name and the main symptom is flaccid paralysis. The kid begins to seem depressed and lies on its side seemingly paralyzed. When moved, it seems as if the kid has no muscles – just floppy. The kid may revive itself, seem normal, only to flop over again within minutes. The other symptom is a distended bowel.

FKS has no other symptoms of illness and no diarrhea or fever. It is my contention that FKS is often misdiagnosed when in fact it is a case of Enterotoxaemia. If there is no symptom of flaccid paralysis, I would believe it is a case of Enterotoxaemia. If FKS is caused by the A type bacterium as suggested, it is unlikey that the treatment currently used would have any effect.

Enterotoxemia can occur in kids under two months and in adult goats. In the periacute form sudden death occurs without prior symptoms and most often in kids under two months of age. The kid will just suddenly die, usually with a scream of pain. Generally within the final hour the bowel will look distended, but can at times only appear distended after death.

Acute form usually shows signs of depression, fever, severe abdominal pain, flaccid paralysis and convulsions. Acute form occurs in both adults and kids. Sub acute form occurs in kids older than 2 months and adults. Goats will show signs for days or weeks, may refuse to eat, lose weight, and have severe diarrhea. Chronic Enterotoxaemia occurs mostly in adult goats. The symptoms can be intermittent, disappearing and reappearing over time. Goats will have a dull, blank stare, diarrhea, lack of appetite, and milk production drops.

Prevention

CD vaccine is the only prevention for the Clostridium perfringens type C and D. There is currently a type A vaccine for cattle, but it has not been tested in goats. The CD/T 8way vaccine should be used and given to kids at 2-3 weeks, then again three weeks later and annually thereafter. Does should be given a CD/T booster three weeks prior to kidding. Any goats previously not vaccinated should receive the initial shot and second dosage as with kids, and then annually thereafter.

Treatment

Stop all milk and feed only electrolytes with baking soda(1tsp bs per 8ozelectrolyte) for 36 hours (at normal bottle schedule) Get 1-2oz or 60cc into kid immediately wait an hour and repeat Administer 7cc of C&D anti-toxin SQ repeat every 12 hrs Antibiotic Tetracycline orally 5-10mg/lb for 5 days.

If goat has no bowel movement or has compacted bowels, use Milk of magnesia 5cc/20lb and do a soft enema.

If goat has diarrhea, give pepto bismal, banamine, and probios.

After bowels return to normal or 36 hrs (whichever occurs first) ease kid back into milk by halving with electrolytes following bottle schedule.

Listeriosis, Circling Disease

First let me say, Listeriosis and Goat Polio are two of the most often confused diseases in goats. Their causes are different; their treatments are different; but symptoms are similar with improper feeding being a major contributing factor of both diseases. <u>Listeriosis occurs most often in adult goats over six months of age, but can occur in goats of any age.</u>

Listeriosis caused by the Listeria monocytogenes bacteria has two forms – Encephalitic and Septicemic. It is more prevalent in the spring and winter months. Listeriosis in either form is a life-threatening disease and is more difficult to treat than Goat Polio.

<u>Humans can become infected with Listeriosis</u> and is associated with the consumption of contaminated meat products, as well as milk and cheese obtained from milk. Humans can also contract Listeriosis by handling fetuses and specimens from aborted animals, and newborns of infected does. Always take appropriate safety precautions when dealing with a goat suspected of Listeriosis.

Listeriosis is usually contracted by ingestion of contaminated water or feed, or by fecal shedding and transfer of the bacteria. Infection may also occur by inhalation and is transferable through the milk of a doe. Excretion in milk is usually intermittent but may persist for many months. Infected milk is a hazard because the organism may survive certain forms of pasteurization. The bacterium of the encephalitic form once ingested migrates quickly to the brain and causes inflammation, while in the septicemic form the bacterium spreads into the bloodstream and infects organs. Infected animals can die within 24-48 hours if treatment is not begun quickly or if improperly treated.

Symptoms

Symptoms usually will appear 10-14 days after initial contact with bacteria. Septicemic is seen more in monogastric species such as Humans and Swine and is not usually a strong culprit within a goat herd. Symptoms most common are diarrhea, abortion, stillbirths, and death. Always wear gloves when handling fetuses and material from aborted does, or newborns from an infected doe. Pregnant women should not handle fetuses or aborted material due to the possibility of contagion.

Encephalitic form is usually seen in ruminants and is also considered to be a sexually transmitted disease in goats. (Italicized symptoms are those that differ from Goat Polio) but if in doubt, treat for both illnesses.

- *may seem depressed or disoriented*

- *may lean against stationary objects*

- *feet walk to the side while goat is walking forward (seen most often in infected adult goats)*

- propel themselves into corners or stand with head against walls, trees or fences

- have a loose or floppy lip

- have a decreased appetite

- have decreased milk production

- may have a fever

- *circle in one direction*

- head tilting to the flank

- have seizures

- *show signs of facial nerve paralysis (on one side)*

- *lack of menace response or ability to close eye on side of paralysis*

- *ear may droop on side of paralysis (seen most often in infected kids)*

- *increased continuous salivation or drooling (seen most often in infected kids)*

- - inability to open mouth or chew cud/cud stuck in mouth (seen most often in infected kids)

- slack jaw

- impaired swelling

- death

Treatment

Recovery depends on early, aggressive antibiotic treatment. If signs of encephalitis are severe, death usually occurs despite treatment. Animals suspected of infection should be isolated from the herd. Sick goats should be taken off grain feed and increase hays and forage browse.

1. Administer Penicillin G at 1cc/15# SQ or 6cc/100# SQ every six hours until 24hrs after all symptoms have disappeared.

   o Follow with tetracycline orally at 11.5 mg/lb per day for 3 consecutive days.

2. Administer adult goats over six months in age Day 1: 6cc/100lb Dexamethasone IM; Day 2- 5cc; Day 3- 4cc; Day 4 – 3cc; Day 5 – 2cc; Day 6 – 1cc. May cause abortions in does.

3. Intravenous fluid/tube feeding electrolyte therapy is recommended.

   o If a doe has Listeriosis, feed kids pasteurized colostrums, or cow's milk.

4. B1 Thiamine and B12 at 5cc/100# orally 1xd for two weeks

5. Add Probiotic Power to feed after symptoms are improving and goat is eating again.

Prevention

Discard spoiled/moldy feed and hay. Maintain proper sanitation of pens, water supply, pasture, and housing. Feeders should be maintained off the ground at no less than chin height of goat. No sudden changes in feed or diet. Wild birds may act as carriers for the disease, try to keep them away from the herd as much as possible.

Goat Polio, Polioencephalomalacia, Cerebrocortical Necrosis

Goat Polio occurs most often in kids being weaned and very young goats under six months of age. In winter, higher numbers of Goat Polio are seen due to the low availability of natural browse and quality of hay available. If ever in doubt between Listeriosis and Polio, treat for both.

Goat Polio is caused by a thiamine B1 deficiency. Changes in the rumen suppress the normal flow of bacteria and interfere with thiamine absorption. Once thiamine is depleted or altered, brain cells begin to die and neurological symptoms appear. Sudden changes in feed, feeding too much grain and not enough hay or browse, moldy hay or feed, use of Corid in treatment of Coccidia, and the usage of antibiotics are all causes that may create a thiamine deficiency in a goat.

Symptoms

(Italicized symptoms are those that differ from Listeriosis) but if in doubt treat for both illnesses.

- appear dull and depressed

- may lie down more often

- may stop eating or drinking

- *Stargazing – may throw their head backwards and up*

- have convulsions occurring 2-5 mins apart

- unable to coordinate muscular movement or muscular contractions

- *may show increased aggression or excitability*

- have muscle tremors

- may have high fever at end stage

- *increased respiratory and pulse rates*

- *have temporary blindness*

- *have severe arching of back*

- *grind their teeth*

- *have severe rigidity*

- *have rapid movement of eyes*

- may stagger or weave

- *have diarrhea*

Treatment

B1 – Thiamine – is the only effective treatment. Most goat B1 meds are at a rate of 100mg per ml/cc. For young goats dosage would be 1/2cc or 50mg per 10# animal every six hours. Adult dosage rates 5cc per 100# which is 500mg per 100# animal. Continue treating until symptoms improve. You can substitute over-the-counter B1 tablets; adjust dosage rates according to mg of B1. It's a lot cheaper and easier to get OTC., and treatment can result in improvement within a few hours if the disease is caught early enough. Overdose with B1 rarely occurs because it is excreted through the body very quickly. If symptoms show no improvement after two treatments, increase dosage to every 4 hours.

Also treat with Probiotic Power at 3-5 TBS as a top feed to encourage normal rumen development. In severe cases, very weak goats or goats not eating, use a Magic drench with above treatment. Dosing with Dexamethasone 0.5 to 1/0mg/lb IM or SC, may help.

Prevention

Monitor feed ration and encourage feeding of more browse and hay. Check incoming hay for mold on a weekly basis. Watch goats for 24hours following use of Corid, dewormers and antibiotics.

Caseous Lymphadenitis, CLA, Cheesy Gland, Lympho, Pseudotuberculosis

CLA is a chronic, contagious disease caused by bacteria Corynbacterium pseudotuberculosis. The bacterium enters a goat's body through mucous membranes or through cuts and abrasions. Rarely, but possible, CLA can be transferred by flies who have come into contact with an infected goat.

The goat's immune system will try to localize the bacteria by surrounding it with cysts. These abscesses are attached to the outside of skin, not the goat's body. When these abscesses burst, the bacterium is spread. The bacteria can survive in shaded areas, on fence posts, and in barns for several months. If there are internal abscesses the bacterium can also be found in feces, surviving in straw, hay, and wood for several weeks.

Symptoms

Cysts or abscesses of CLA do not usually show up until 2-6 months or more after the initial infection. If your goat has CLA, you will notice lumps, cysts, or abscesses near the jaw, in front of a shoulder, under an ear, or near a doe's udder attachment. The goat may also develop internal cysts. Any abscess located near a lymph gland should be considered CLA.

A blood test can determine if a goat has CLA, but the most effective test is of the pus from the cyst itself. Pus from CLA is thick, pale green, dry, cheesy looking and highly infectious. Other non-CLA abscesses have a thinner liquid pus and foul smell.

Prevention

Prevention is the key by developing strict quarantine measures for new goats, maintaining clean housing, sterilizing tools prior to use, practice fly control, and not bringing infected goats into your herd.

Treatment

Always wear rubber gloves when treating CLA, preferable ones that come up to the elbow. Currently there is no verifiable cure for CLA, and the only vaccine available is not to be used in goats due to the extreme side effects shown during testing. Antibiotics cannot treat CLA.

When CLA is noticed in a herd, you should remove the infected goats from the herd and cull them. If you choose not to cull infected goats, isolate the sick goat from the herd, drain abscesses and disinfect and cover open wounds. All areas where the infected goat had contact, and all equipment, should either be burned or disinfected with a full bleach solution. This will need to be done on an ongoing basis as eliminating the virus contagion (draining abscesses) does not eliminate the virus from the goat.

CAE (Caprine Arthritis Encephalitis)

Please look at our disease review chart because so many of the goat diseases have similar symptoms and CAE can be mistakenly diagnosed. All goats can be infected, however CAE is very uncommon in meat/fiber goats such as the Pygmy goat. CAE is an infection for life as there is no current cure for the disease. Only 30% of infected goats show clinical symptoms. (Caprine Arthritis and Encephalitis, Small Ruminant Lentivius Infection Content Update: March 15, 2007)

The Encephalitis Form of CAE

This form of CAE is a virus which causes inflammation of the spinal cord and destroys nerves controlling the motor function of hind limbs. In kids, signs usually begin between one and six months of age. Your goat would show signs of lameness, lack of coordination, or weakness in one or both rear legs. Nerves which control motor function of the hind limbs are progressively destroyed. Over the course of days to weeks, the weakness usually progresses to paralysis.

The young goat will usually remain bright and alert and continue to eat and drink in the beginning. Mild pneumonia may develop. Encephalitis CAE in older Pygmy and Nigerian goats is rare but mimic symptoms of Listeriosis and Tetanus. Signs include circling, head tilt and facial nerve paralysis.

The Arthritic Form of CAE

The arthritic form of CAE usually appears between one and two years of age. Gradually developing lameness accompanied or followed by swelling of the joints is usually the first outward sign. Swelling is most often noted in the front knees and can also be seen in the hock and stifle joints. As it progresses in goats, joint pain and stiffness become more apparent with the goat spending most of its time lying down or having to walk on its knees. The goat will begin to lose weight and develop a rough hair coat. A doe's udders will harden and produce less milk or stop producing milk at all. Pneumonia can be a symptomatic reaction of CAE. As with the Encephalitis CAE there is no cure for Arthritic CAE. The goat can be made more comfortable by following proper goat health, foot care, anti-inflammatory medicines, and proper diet.

Transmission, Diagnosis and CAE Cure

CAE is transmitted by body secretions which contain infected white blood cells. Infected mothers pass CAE to their kid through colostrums, body fluids or milk. Other bodily fluids, such as blood, open wounds, saliva , and semen have shown to carry the CAE virus. Diagnosis is usually through goat history and symptoms. However there are blood tests that determine if a goat has the CAE antibodies. This test however does not confirm the goat has the disease. Positive antibodies can mean a goat has a natural immunity to the CAE virus. There may also be false negatives when a goat does in fact have CAE. There is no known cure for CAE.

Johne's (pronounced "Yoh-nees") disease, Paratuberculosis

Johne's disease is caused by a bacterium named Mycobacterium avium ss. paratuberculosis (MAP). Goats become infected in utero or within the first few months of life. However, infected goats remain healthy, active and alert for many months to years prior to developing symptoms.

The bacteria will lay quiet within the ileum of the small intestine and then suddenly begin to take over more and more tissue, causing inflammation which thickens the intestinal wall causing abnormal function. It is contagious and can be spread species to species – from goat to goat, cow to goat, goat to cow, etc. Although it is interspecies transferrable, it is suspected but has not yet been proven to be a cause of Crohn's disease in humans. Johne's is a very hardy disease and is resistant to heat, cold and dry weather. It cannot replicate outside of an infected animal.

Most often, Johne's disease is brought into a herd by an infected animal. The infected goat will then shed the bacterium through fecal material onto the property. Young goats are more susceptible to infection than are adults. Goats will ingest the bacterium when consuming infected plants, grass, hay, or water. Other possible transmissions occur from bottle feeding infected milk to kids or kids born from an infected mother.

Symptoms

An infected goat will show rapid weight loss and diarrhea. But diarrhea is less common. Goats will continue to eat well, but will lose weight fast and become weak. Because these are the only symptoms of Johne's disease, it is very easy to misdiagnose. Laboratory tests are the only way to confirm diagnosis of the disease. Because of the high contagion factor, if a goat is found to be infected all goats in the herd should be presumed infected as well.

Prevention

The only way to prevent this disease is to not bring infected animals into your herd. Because of the long timeframe before symptoms are seen, quarantine procedures will not rule out an infected goat. Purchase goats only from Johne's free herds. Testing can be done on a herd through fecal sample cultures, by blood or milk antibody testing.

Treatment

A goat or herd with Johne's disease should be culled. There is no current vaccine for Johne's disease. Antibiotics will not help treat the disease. Studies have shown giving goats multiple antibiotics daily for several months have subdued the symptoms but they reappear once treatment is stopped. There have been few studies for the efficacy of antimicrobial drug treatment for the MAP bacterium in animals or in humans.

Pregnancy related Disorders - Chlamydia

There are several variant strains of Chlamydia, however the most common one affecting goats is Chlamydia Psittaci. It is a zoonotic bacterial infection which can spread interspecies – birds to goats, goats to human, etc. When there are a lot of abortions within a herd, Chlamydia is the usual culprit.

<u>Transmission</u>

- Infected animals excrete large amounts of Chlamydiae in the placenta and fetal fluids at the time of parturition and at the time of abortions.

- Goats may shed Chlamydia in vaginal fluids from two weeks prior to abortion to two weeks after the abortion.

- Smaller amounts can also be shed in urine, milk, and feces several days after the abortion

- Animals can pick up the disease by inhaling the particles through feed, water, or dust particles.

- If they are 100 days pregnant they are more susceptible to the disease than one at the end of gestation or barren.

- Infected mothers can pass the disease onto their young which causes it to stay in the flock or transmit it to others.

- It can be transmitted to does through the direct contact of feces from infected pigeons and sparrows.

- Chlamydia can be also transmitted to goats by ticks or other bloodsucking insects.

Symptoms

Pregnant does may, without other symptoms experience:
- abortions – usually 60-90% of herd does
- premature delivery, usually late gestation
- mummified babies
- weak or stillborn babies
- afterbirth retention

Adult goats may experience:
- persistent cough without breathlessness
- arthritis
- keratoconjunctivitis

Treatment

Always wear gloves and use protective measures when handling infected animals/embryo's/dead kids/amniotic sac as it can in rare cases be transmitted to a human.

Pregnant Does after 30 days pregnant – Biomycin 1 cc per 20 lb given once weekly , SQ,until the doe kids.

Bucks and kids over 6 months – LA200 or Oxytetracycline 3ml/100# once daily for three days.

Pink Eye Infection – two drops of Oxytetracycline directly into the eye for three days.

Prevention

Good news is there is a vaccine which can be given to prevent abortion outbreaks in the future. However, all goats must be vaccinated and even with use of a vaccine, a herd already infected may take up to 3 years for abortions to stop. Killed vaccines can prevent the abortions, but cannot prevent the shedding of Chlamydia bacteria at kidding. Follow sanitary measures during and after birthing.

Clean Up

Placentas, bodily discharge material, and dead babies should be burned. The housing/birthing area should be cleaned out with a highly potent disinfectant such as a phenolic like Pine II or a quaternary ammonium compound similar to SaniZide.

Toxoplasmosis, T Gondii, Toxo

Toxoplasmosis is a parasite infection caused by Toxoplama gondii. It is zoonotic which can spread interspecies – cats to goats, goats to human, etc. The initial transmission is from inhalation or consumption of cat feces infected with the oocysts. The parasite undergoes various cycles within the intestinal tract of a cat which then excretes millions of oocysts in its feces. These oocysts then sporulate within one to five days. They may survive for several months before finding a new host for infection.

All warm-blooded species can be infected. Within one to two weeks after ingestion, the parasites infect brain, muscle, and placenta of pregnant mammals causing the immune system response to slow and formation of cysts. Once infected with toxoplasmosis, the goat builds immunity and will not become infected again. Humans can become infected by eating infected animals as well as by consuming milk from an infected animal.

Symptoms

General symptoms may or may not be present which include fever, diarrhea, cough, shortness of breath, jaundice, and seizures.

Infection in non-pregnant does and early gestation

- Infertility
- Death and reabsorbtion of fetus

Infection in mid gestation

- Abortion
- Stillbirths
- Weak kids
- Kids die within a few weeks of birth

Infection late gestation

- Live birth of infected but immune kids

Prevention

Controlling or eliminating cat population within a herd; Infection of herd prior to pregnancy through contaminated ingestion; Use of a vaccine currently used in sheep.

Treatment:

It is difficult to deter an abortion outbreak caused by toxoplasmosis once abortions begin. However, some effective measures have been obtained by use of prescription Decoquinate at 2mg/kg bw/day throughout pregnancy or a combination of sulphamezathine at 15-25 mg/kg with pyrimethamine at 0.44 mg/kg before abortion occurs.

The diseases and illnesses discussed in this chapter are not the only ones goats may get or come in contact with. Those listed are the ones I believe to be most prominent. Do some research and have a qualified, goat veterinarian available before you ever need one.

Chapter Ten – Antibiotics

(The following is for informational purposes only. Most drugs are used off-label on goats. Please always check with your local veterinarian for administration and dosages)

SubQ injections are injection given under the skin. It is easiest to pick up the skin right behind the front shoulder, over the ribs. Pull out the skin upward forming a tent. Insert the needle into the tent from the top side of the goat, pointing the needle towards the ground.

IM injections are given in the muscle of a goat and are not recommended by many veterinarians. If needed, it is suggested to give only the first shot IM and then subsequent shots for dosing SubQ. When you think about it, it makes sense, if shots are needed for five days or longer, what muscle will be left to use that hasn't already been done?

Needle gauges – We use 18 gauge ½ half inch needles on viscous liquids like Penicillin or Ivomec. For thinner liquid injections use a 20 gauge needle at 1/2 inch long.

Antibiotics are becoming more and more overused in goat Health. Most times, if the goat is not running a fever an antibiotic is not necessary. Overuse of antibiotics makes them less effective for proper, accurate treatment. Please don't use antibiotics as a first line treatment when something goes wrong.

Antibiotics – Always give antibiotics for a full five day treatment. Anything less will encourage bacteria resistance. Also, using antibiotics as a general treatment when not needed encourages resistance. Antibiotics cause goat stress and negatively affect the goat's digestive track. We only treat with antibiotics when there is fever present and always use probiotics when on antibiotic treatment.

Listed Alphabetically – ALL dosage for SubQ, 5 day treatment unless specified otherwise.

Brand Name	Drug Name	Treatment	Dosage	Milk Withhold	Note
Agri-Cillan, Pfi-pen, Us Vet Penicillin	Procaine Penicillin G (300,000 per ml)	streptococcus infections, chronic pneumonia and other infections	1ml/15# 2xd until symptoms disappear but at least 5 days	14-20 days	Do not use with Oxytetracycline as it makes ineffective
Agromycin, Biomycin, Geomycin, LA200, Liquamycin, Maxim, Oxytet 200,	Oxytetracycline-200 mg/ml	keratoconjunctivitis ("pinkeye"), mycoplasma & Chlamydia, metritis, mastitis	3ml/100# once daily for 2 days; for pinkeye – also dose 3-4 drops in eyes.	12-18 days	This is NOT Tylan 200; Milk makes ineffective; Do NOT use on pregnant or under 6mo.
Bactrim, Cotrim, Ditrim, Septra, TMP	Trimethoprim/ Sulfaethoxazole Trimethoprim/ Sulfaiazine	scours, pneumonia, UTI, Bladder infections, and other infections	(1)960mg tab/70# 2xd	8 days	

Brand Name	Drug Name	Treatment	Dosage	Milk Withhold	Note
Crystiben, Pen BP48, Twinpen, US Vet Penicillin Benzathine	Penicillin Procaine & Penicillin Benzathine combo(150,000 ea per ml)	Streptococcus infections, chronic pneumonia and other infections.	1ml/25# 1xd until symptoms disappear but at least 5 days	25-30 days	Do not use with Oxytetracycline as it makes ineffective
Excenel	Ceftiofur hydrochloride	Respiratory; foot rot	1ml/25# 1xd	None	2nd gen Penicillan
Naxel	Ceftiofur sodium 1mg/ml	Respiratory	1ml/50# 2xd	None	5dy shelf life; use with other antibiotic
Nuflor	Florfenicol-300mg/ml	salmonellae and E-coli	3ml/50# 1xd	28 days	Very Painful injection
Today, Cefalac	Cephapirin Sodium	Mastitis	4-5 infusions/ 12hrs	6 days	Milk out completely prior to treatment, milk often after treatment
Tomorrow, Cafadri	Cephapirin Sodium(long acting)	Mastitis	1 tube per udder		Use on Dry does
Tylan 200	Tylosin 200	mycoplasma, chlamydia, rickettsia, upper respiratory infections, enteritis, mycoplasma arthritis in kids	1ml/20# 1xd	8 days	Very Painful injection

Chapter Eleven – General Medicines

(The following is for informational purposes only. Please

always check with your local veterinarian for administration and

dosages).

Albon – *see also Dimethox 12.5% liquid * Five day

treatment to treat Coccidiosis- you must treat the full five days.

Day one: 1 ml per 5 pounds- given orally. Days 2-5: 1 ml per 10

pounds- given orally.

Apple Cider Vinegar (ACV) – place in water to help acidify

the urine and prevent formation of calculi in bucks and will also

help decrease algae in the container. 2cups per 5 gallons water

weekly.

Aspirin – used for fever and inflammation. Dose 325mg (1

adult aspirin) per 10 lbs. Milk withhold – 24hrs

Banamine (Fluxixin Meglumine- 50mg/ml) – Always take

temp prior to giving; if temp is low do NOT give. It is used for

fever, smooth muscle relaxant, pain reliever and to stimulate the

rumen. Dose 1 cc per 100 lbs body weight IM, but can be used at a rate of 1/2 cc per 25-30 lbs once daily - treat no more than three days. Keep refrigerated. Milk withhold – 3-4 days.

Baking soda – Leave out free choice for regular rumen care. Also for emergency treatment for bloat. Dose – 1 Tbsp in 30cc water as drench.

Benadryl – Antihistamine, for minor allergic reactions such as bites, coughs, and nasal decongestant. Over the counter dose 1tsp/ 5cc for kids and 2-4 tsp/10-20cc for adults.

Beer (dark malt is best)- It is used for bloat. Dose = one can.

Betadine solution -for wounds; apply externally.

Blue cohosh – contains uterine contracting material, and works like oxytocin without the harmful effect. As a tincture form dose directly on the tongue, 15-20 drops 3-4 times a day.

Bo-se – given for selenium deficiency; dose 1cc/40lbs twice a year SubQ. We do not use this for kids as an overdose can kill. Milk withhold – 24hrs.

Calcium gluconate 10% (100mg/ml) It is used to treat milk fever, pregnancy toxemia, and ketosis. This must be given IV only at 5-15mg during a 10min span. It is caustic to tissue if given IM or sq.

Charcoal Fish aquarium – 4-6 oz to treat poisonings. It binds with poison to remove from system. Dose as a top off to feed or mixed with water or liquid laxative. If used with water or as top feed, follow with a laxative.

Chlortrimeton – OTC for colds and coughs similar to Benadryl without drowsy effect. Dose– 5cc for kids and 10-20cc for adults.

Corral and Sevin dust– can be used on non lactating does and bucks only – do not use on kids. For adults sprinkle on and rub in. Apply second dose 10 days later.

Dimetapp -OTC for colds and coughs. Dose 3-5cc for newborns – 2wks in age; 4-6 tsps for adults

Dimethox 12.5% liquid – *see also Albon* – To treat a herd that is already infected with coccidian. Dose administer 3 to 5 cc's of undiluted liquid Di-Methox 12.5% orally to each kid daily for five consecutive days; adults 8 to 10 cc's. Preventative dosage is usually one-half the curative dose; read product labels.

Dopram – It eliminates respiratory distress in newborns caused by troubled births, and to stimulate lung activity. Dosage is to drop 2/10 cc under kid's tongue immediately upon birth. Keep refrigerated.

Epinephrine – Used as a counter shock. A MUST-Have for every goat medicine cabinet. Dose 1 cc SQ per 100 pounds body weight.

Enemas – Useful for constipation and toxicity reactions, including Floppy Kid Syndrome. Other uses: If you have a baby girl born with vagina turned out, use a children's enema rectally to move her bowels and the vagina will most times return to normal position.

Formalin (10% buffered formaldehyde) – This is used off label to treat CL and hoof rot.

Gas X (or generic brands) – used for bloat. Dose 1-2 tablets every 2-4 hours; Liquid form 1-2 capfuls.

Geritol – It is used for anemia. Dose 1Tbsp every day for 2-3 days.

Ginger- To sooth stomach and intestines. Dose 2Tsp.

Ibuprofen – Anti-inflammatory, fever reducer, pain killer. Double human dose. Milk withhold – 24hrs.

Kitten flea and tick powder used for lice treatment in kids. Sprinkle from head to tail and rub in. Dose once and then again in 10days.

Kopertox – OTC product used to treat hoof rot and hoof scald. Dose topically as needed.

Listerine – It is used to treat ear mites and to clean the ear. Dose 1 capful – dip qtips in solution and cleanse inside of ear, wipe out with tissue.

Milk of Magnesia – Useful for constipation and toxicity reactions, including Floppy Kid Syndrome. Dose as oral drench 15 cc per 60 lbs.

Mineral Oil – Useful for ear mites by placing 2-4 drops in ear canal.

Molasses/Karo Syrup – Useful for ketosis in does and as an energy boost. Dose at 5-6 cc orally twice a day for ketosis.

Neosporin (or generic) – antibiotic ointment . Used for scrapes and small wounds. Apply as you would on a human.

Nuflor (Florfenicol) – Used to treat pneumonia. Administer IM every other day for a maximum of three injections. Use Luer Lock syringes. Dosage is 1 cc per 25 lbs. of body weight, no more than 3cc per 100lbs. Keep refrigerated.

Pepto-Bismol – Helps control diarrhea in kids under one month old. Use up to 2 cc every four to six hours for newborns; 5 cc for kids approaching one month old. Follow up with Probiotic Power

Probiotic Power or generic – Used to return normal flora in gut. Dosage for sickly goats or goats just not up to par is 3 to 5 Tbs as a top feed.

Redcell- Is used to treat anemia and as a vitamin /iron supplement. Dose 6cc per 100lbs orally.

Robitussin – Used for coughs and colds. Dose 3-5cc for newborns – 2wks in age; 4-6 tsps for adults.

Slippery elm – Helps stop diarrhea and coats the stomach and intestines. Dose 1 tsp mixed with yogurt or probiotics and electrolytes so it can be drenched. Give 5-6 times over the day until diarrhea stops.

Sulfaquinoxaline 20% as a drench- Used to treat Coccidiosis. Dose 2ml/50lbs by mouth for 5 days.

Sulmet 12.5% – Prevents or treats Coccidiosis. Dose as prevention 1cc of solution per 5lbs of goats day 1 and 1 cc per 10 lbs of goats for day 2 thru 5 as a prevention (spring and fall each year); as treatment you must treat a full five days. Day one: 1cc/5lbs; Days 2-5: 1cc/10lbs given orally.

Tagamet- Used to treat Coccidiosis. Dose- One 200mg tablet given 1 a day for 3-5 days; kids get ½ tablet.

Tavist D – Treats colds and coughs. Dose 3-5cc for newborns – 2wks in age; 4-6 tsps for adults.

To-Day (cephapirin sodium) – Over-the-counter product for mastitis treatment. Milk out the bad milk/pus/blood and infuse one tube of To-Day into each infected udder for a minimum of two consecutive days.

Tylan 200 (tylosin) – Used to treat respiratory problems. Use 1 cc per 25 lbs. body weight for five consecutive days intramuscularly (IM). Keep refrigerated.

Vitamin C – Helps acidify the urine and relieve congested udder. Dose urine acidifier – 250mg tablet 4x day; congested udder – five 500mg tablets 2x day.

Vitamin E – Works with selenium and is essential for tissue, muscular and udder health. Used as a supplement before kidding. Dose 500 to 1000 iu as top feed daily.

B1 and B12

I use over the counter human consumption vitamins you buy at grocery stores and never have had a problem with it. I worked out the dosage years ago which was pretty simple. B1 and B12 is so important to a goat's health, it is a MUST for every medicine cabinet. We give it to all new herd mates their first week and anytime a goat is sick or not acting normal.

My experience with a sick goat – I had a goat for six months who all of a sudden just started acting 'off'. Separated from the herd, the goat would lay down a lot and was off feed. He would go to the fence or a tree and stand there with his head against it – kind of like a child in a corner. Normally my first thought would have been Listeriosis. But, I did a thorough checkup and found no other symptoms or issues. After 24hrs of watching him with no improvement or changes, I got him in our sick pen and started him on the Magic Ball recipe (at the end of this chapter). Within one day of treatment, he started acting normal again. IMO most issues in goats can be resolved with B1/B12 treatment.

B1 Thiamin

Provides energy by converting blood sugar into energy. It keeps mucous membranes healthy and is essential for nervous system, cardiovascular and muscular function. A lack of thiamin in the goat's system can be devastating and may become fatal. Goats produce their own thiamin but when there is a thiamin deficiency the result can be swelling in the brain, temporary

blindness, incoordination and staggering (Goat Polio-

Polioencephalomalacia) which can result in death. Treatment

with B1 can show drastic improvement within a few hours.

B12 cobalamins

Keeps nerves and red blood cells healthy. It is responsible for

the smooth functioning of several critical body processes. It

converts carbohydrates into glucose leading to energy

production and a decrease in fatigue and lethargy. B12 helps in

healthy regulation of the nervous system, reducing depression,

stress, and brain shrinkage, and it helps maintain a healthy

digestive system. It is essential for healthy skin, hair, and

hooves. It also helps in cell reproduction and constant renewal

of the skin.

When a goat stops eating or when the digestive system is not

able to absorb this vitamin well (usually from bacteria growth in

the small intestine, or a parasite), a deficiency in vitamin B12

occurs. B12 deficiency can be fatal in goats and long-term

treatment may be necessary. B12 deficiency results in illnesses

like anemia, fatigue, weakness, constipation, loss of appetite, weight loss, depression, poor memory, soreness of the mouth, vision problems, and a low sperm count.

Vitamin B1 Thiamine – Used to treat goat polio, rumen productivity and provides increased energy. Dose—250mg B1 tablets – 2 tablets the first hour, 1 tablet in six hours after first dose, then 1 tablet twice a day for five days.

Vitamin B12 Cobalt – Treats listeria, anemia , stress, and as an appetite stimulant; Dose – 2000mcg B12 tablets – 2 tablets the first hour, 1 tablet in six hours after first dose, then 1 tablet twice a day for five days. Keep refrigerated.

Yogurt Plain – to help replace the normal flora in the rumen after rumen problems like bloat and after giving antibiotics. Dose–give about 1/4 cup 2-3 times a day for 3-5 days.

Best way to get goat to eat pills – place index finger and thumb on each side of goat mouth near back and open mouth. Take other hand and place tablet in very back of throat but Not down throat. Cup the hand you were using to open mouth around sides of mouth to keep goat from spitting it out, but not tight enough so he can't chew tab. Repeat for each

tablet. One dosage usually takes about 1-2mins.

As the B vitamins go through the system pretty fast, I haven't

worried about overdosing and it has always worked for me. No

needles, no frustration over needing a prescription or having to

buy Fortified B complexes for goats when I just need the B1 or

B12.

When a goat is acting off or needs a pick-me-up, Give him
some **Magic**.

Magic is used as a drench or can be made into balls. It is
Essential to give to your goat whenever they are sick or just not
acting right. If using the drench method, provide goats with top
feed of probiotics.

MAGIC Drench (recipe makes treatment dosage for five
days)

1 part corn syrup

1 part corn oil

1 part molasses

5 tablets of 250mg Vitamin B1

5 tablets of 2000mcg Vitamin B12

We use an ice cream scoop to measure corn syrup, corn oil,
and molasses into a bowl at two scoops each. Grind B1 and
B12 in blender, and then add to bowl. Stir well and store in

container until needed. If stored in refrigerator, it may become too thick to dose well. Just allow to set at room temp for about an hour. Dose two tablespoons 2x day. May use drench syringe for dosing.

MAGIC BALLS

1 part corn syrup

1 part corn oil

1 part molasses

5 tablets of 250mg Vitamin B1

5 tablets of 2000mcg Vitamin B12

5 scoops of Probiotic Power probios

2 cups of alfalfa feed pellets

 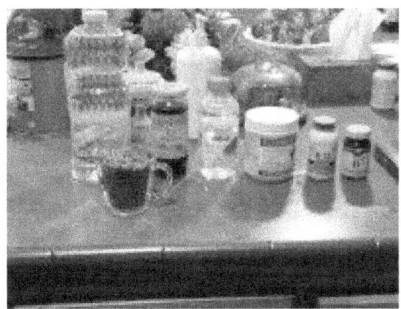

We use an ice cream scoop to measure corn syrup, corn oil, molasses into a bowl at two scoops each. Add in Probiotic. Grind B1, B12, and feed in blender, then add to bowl. Stir well, form into golf size balls. If hands become too sticky to make balls, wash hands and try again. Wrap balls individually in plastic wrap and store in refrigerator until needed. Allow to defrost one hour prior to use. Makes approx. 5

balls. Dose one ball per day for five days.

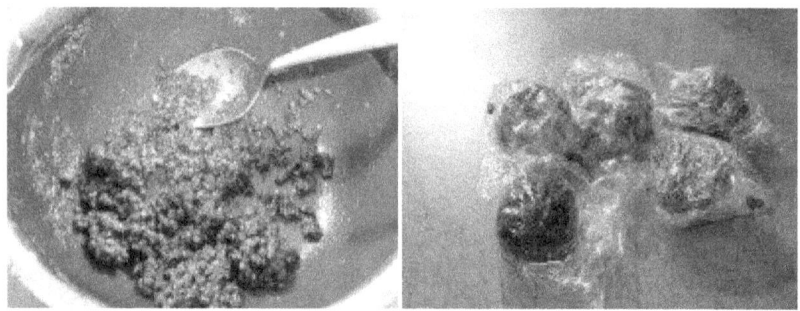

If you start now by preparing a Herd Management program and keep your goats healthy, you will have much more success in raising goats than you would without a plan. And if you can, find a mentor in your area, someone who knows a lot about goats and is willing to help you as you learn what you need to know. Just remember – Goats are addictive! I suspect you will end up with more than you plan. I wish you all good fortune and success in your endeavor.

Glossary

Common words, abbreviations, and terms explained.

AC – Ammonium Chloride

ACV – Apple Cider Vinegar

BCS – Body Condition Score

BOSE – a selenium mineral supplement

BOSS – Black Oil Sunflower Seeds

Buck – sexually intact male

CAE – Caprine Arthritic Encephalitis

CD/T – a vaccine for enterotoxaemia C&D and for Tetanus

CL – either Caseous Lymphadenitis or Craigs List

Dehorned – horns removed after growth has occurred

Disbudded – horn buds removed before horn growth

Doe - female

FF – First Freshener - A doe that is pregnant or with milk for the first time

Grade – a grading system of health conformation in meat breeds; a goat which has one purebred parent and one mixed parent

Kid – a young goat under six months of age

LGD – Livestock Guardian Dog

Mixed – a goat that has two different breed parents

Nigie – Nigerian goat

Pen G – Penicillin G

Polled – naturally hornless

Probios – Probiotics – good bacteria to help other bacteria in gut digest food

Purebred – a goat that is 100% one breed; no mixed breeds

Pygerian – Pygmy and Nigerian cross

Rumen – Large part of the reticulorumen in the gut used to ferment ingested food

Scrub – a goat with one purebred parent and one unknown breed parent

Scur – horn growth after disbudding

UC – Urinary Calculi – Similar to kidney stones

Wether – male who has been castrated

Reference Material

The following information are websites, books and other reference material that I found useful. I cannot guarantee the links will still be available.

In-General Information

www.elwoodranch.com

www.tennesseemeatgoats.com

www.goat-link.com

www.fiascofarm.com

http://www.luresext.edu/goats/training/qa.html

www.goatworld.com

Medicine Conversion

http://tinyurl.com/82yznhb

http://tinyurl.com/bnno6p7

Dairy Goats

http://tinyurl.com/dzoqvp

http://cru.cahe.wsu.edu/cepublications/em4894/em4894.pdf

http://tinyurl.com/bqq8nru

Dietary

http://tinyurl.com/d49zw38

http://idgr.info/index/articles/the-feeding-of-goats/

http://www.goatspots.com/copper.html

Diertary cont.

http://www.gov.mb.ca/agriculture/livestock/goat/pdf/bta01s08.pdf

http://www.guinealynx.info/hay_chart.html

Diseases

http://tinyurl.com/7pwfrjl

http://www.merckmanuals.com/vet/index.html

http://tinyurl.com/bwl5woq

Drawing Blood

http://www.boergoats.com/clean/articleads.php?art=64

Symptom Chart

http://www.elwoodranch.com/goats/goat-care/symptom-finder/

Fecals

http://tinyurl.com/cvzolgt

Meat Goats

http://www.luresext.edu/goats/training/marketing.pdf

http://www.ncagr.gov/markets/mktnews/goatgrades02.pdf

Parasites

http://tinyurl.com/bm5y6cu

http://www.gentlecarepet.com/mange_mites_fungus

http://tinyurl.com/c6rdsdk

Pregnancy Care

http://kinne.net/cseccare.htm

http://tinyurl.com/cmreoz2

http://tinyurl.com/cssytd8

http://tinyurl.com/bqvnbzv

http://tinyurl.com/c4sf78d

http://tinyurl.com/cz9lpnb

Supply stores

www.tractorsupply.com

www.hoeggerfarmyard.com

www.jefferspet.com

Tube feeding

http://tinyurl.com/boyvkr5

http://tinyurl.com/ccm2gsr

www.ingramcontent.com/pod-product-compliance
Lightning Source LLC
Chambersburg PA
CBHW081451170526
45166CB00008B/2393